ENVIRONMENTAL REMEDIATION TECHNOLOGIES, REGULATIONS AND SAFETY

FEDERAL CONTROL OF AIR EMISSIONS

SELECT TYPES OF CONSIDERATIONS

ENVIRONMENTAL REMEDIATION TECHNOLOGIES, REGULATIONS AND SAFETY

Additional books in this series can be found on Nova's website under the Series tab.

Additional E-books in this series can be found on Nova's website under the E-book tab.

ENVIRONMENTAL SCIENCE, ENGINEERING AND TECHNOLOGY

Additional books in this series can be found on Nova's website under the Series tab.

Additional E-books in this series can be found on Nova's website under the E-book tab.

ENVIRONMENTAL REMEDIATION TECHNOLOGIES,
REGULATIONS AND SAFETY

FEDERAL CONTROL OF AIR EMISSIONS

SELECT TYPES OF CONSIDERATIONS

HARRISON M. GRAHAM
AND
TODD E. LEE
EDITORS

nova publishers
New York

For permission to use material from this book please contact us:
Telephone 631-231-7269; Fax 631-231-8175
Web Site: http://www.novapublishers.com

Additional color graphics may be available in the e-book version of this book.

LIBRARY OF CONGRESS CATALOGING-IN-PUBLICATION DATA

ISBN: 978-1-62081-808-4

Published by Nova Science Publishers, Inc. † New York

CONTENTS

PREFACE

This book examines select types of considerations in the federal control of air emissions. Topics discussed in this compilation include controlling air emissions from outer continental shelf sources; the EPA's Boiler MACT standards to control emissions of hazardous air pollutants; tall smokestacks and their contribution to interstate transport of air pollution; the revisions and associated issues relating to the National Ambient Air Quality Standards for Particulate Matter; and a summary of the Clean Air Act and its major requirements.

Chapter 1 - Air emissions from outer continental shelf (OCS) operations are subject to different regulatory programs, depending on the location of the operation.

The Department of the Interior (DOI) has jurisdiction over OCS sources in federal waters in the western Gulf of Mexico and most of the central Gulf. In addition, the Consolidated Appropriations Act, 2012 (P.L. 112-74), transferred air emission authority in the OCS off Alaska's north coast from the Environmental Protection Agency (EPA) to DOI. EPA has jurisdiction over sources in all other federal waters.

The primary difference between the EPA and DOI programs is rooted in the different statutory authorities: the 1990 Clean Air Act (CAA) and the 1978 Outer Continental Shelf Lands Act (OCSLA). The primary objectives of these statutes are different—air quality versus offshore energy development. The two regulatory programs reflect these underlying differences.

For much of the past 30 years, these differences received little attention, primarily because most of the federal oil and gas resources in EPA's jurisdiction have been subject to moratoria. In 2008, moratoria provisions expired, potentially opening many of the areas in EPA's jurisdiction to oil and gas leasing activity. If more OCS areas in EPA's jurisdiction are open for oil and gas leasing, policymakers interest in these differences will likely increase.

For OCS sources in EPA's jurisdiction, requirements depend on whether the source is located within 25 miles of a state's seaward boundary ("inner OCS sources") or beyond ("outer OCS sources"). Inner OCS sources are subject to the same requirements as comparable onshore emission sources, which vary by state and depend on the area's air quality status; outer sources are subject to various CAA provisions, including the Prevention of Significant Deterioration (PSD) program. In contrast, OCS sources in DOI's jurisdiction are subject to air emission requirements only if emissions would "significantly affect" *onshore* air quality.

A key difference between the EPA and DOI programs is the federal emission threshold that would subject a source to substantive requirements.

For sources in EPA's jurisdiction, this is the PSD threshold of 250 tons per year (tpy) of regulated emissions. Sources that exceed this level would likely be subject to Best Achievable Control Technology (BACT) and other provisions. States' analogous thresholds that apply to inner OCS sources may be more stringent. By comparison, a DOI OCS source applies an exemption formula, based on distance from shore (e.g., a source 30 miles from shore would have an emission threshold of 990 tpy).

If a source remains subject after this step, it must conduct air modeling to assess whether its emissions would have a significant affect on onshore air quality. In effect, this two-step process constitutes a much less stringent threshold than EPA's 250 tpy threshold.

Another substantial difference is the time frame allotted to the agencies for reviewing a potential source's permit (EPA) or activity-specific plan (DOI). In addition, the EPA permit process allows greater opportunity for input from the public. In particular, EPA's Environmental Appeals Board offers parties a powerful tool to compel agency review.

Therefore, two identical operations, located in separate jurisdictions, could face considerably different requirements and procedural time frames. Some stakeholders would likely argue that the additional opportunities for public involvement in EPA's permit process help create a balance between resource development and environmental concerns. Others would likely contend these steps present unnecessary burdens and timing uncertainty in the process.

Chapter 2 - On December 2, 2011, EPA Administrator Lisa Jackson signed proposed revisions to EPA's recently promulgated Maximum Achievable Control Technology standards for boilers (the "Boiler MACT").

Publication of the proposed revisions in the Federal Register on December 23 began a 60-day public comment period, after which EPA will complete reconsideration of the rule—presumably by mid-2012. The Boiler MACT standards now being reconsidered were promulgated March 21, 2011, to meet the requirements of Section 112 of the Clean Air Act, but EPA stayed implementation of the requirements until it completes the reconsideration process.

There is widespread interest in the rule's requirements and their potential effects, because boilers are used as power sources throughout industry and for power or heat by large commercial establishments and institutions as well.

EPA developed the regulations because it has found, based on emissions data, that boilers (including coal-, biomass-, and liquid-fired boilers) are major sources of hazardous air pollutants (HAPs).

The Clean Air Act defines a major source as any facility that emits 10 tons or more of a single listed HAP or 25 tons of any combination of HAPs annually. The HAPs themselves (187 substances) were listed by Congress in the 1990 Clean Air Act Amendments.

When finalized, the rule will replace a 2004 version of the rule that was vacated and remanded to EPA by the D.C. Circuit Court of Appeals. EPA has been under a court order to promulgate a replacement.

As proposed December 2, the MACT would affect about 14,000 boilers and process heaters, with capital costs of $5.4 billion, according to the agency; annualized costs, which spread the costs of capital over the expected life of the equipment and include operating and maintenance expenses, are estimated at slightly less than $1.5 billion per year. In response to comments on an earlier proposal and on the promulgated rule, the current EPA proposal would reduce the number of units expected to require controls, and make the emissions standards much less stringent, reducing the agency's estimate of annualized control costs from $2.9 billion to $1.49 billion.

Most of the costs would be borne by boilers that burn coal, biomass, or liquid fuels; only 12% of all the units covered by the rule will need to install equipment to meet it. Most of the boilers affected by the rule (83%) are fueled by natural gas or refinery gases. These boilers would not have to install pollution control equipment and most would experience cost savings under the rule, according to EPA. For the rule as a whole, EPA estimated that benefits—including the avoidance of 3,100 to 8,000 premature deaths annually—would outweigh costs by at least $25 billion per year.

Affected industries and many in Congress have raised objections to the rule as proposed and as promulgated, and bills have been introduced in both the House and Senate (H.R. 2250 and S. 1392) to alter the rule's requirements and delay its implementation. H.R. 2250 passed the House 275-142 on October 13.

Provisions similar to H.R. 2250 were offered as an amendment (S.Amdt. 1660) to the Senate version of the surface transportation bill (S. 1813) on March 8, 2012, but were not adopted.

In addition to the Boiler MACT, this report discusses three related rules that EPA promulgated at the same time, dealing with smaller "area source" boilers and with commercial and industrial boilers that burn solid waste (the "CISWI" and solid waste rules).

The latter two rules have also been controversial. Like the Boiler MACT, the CISWI rule was stayed by EPA, pending reconsideration.

Chapter 3 - Tall smokestacks—stacks of 500 feet or higher, which are primarily used at coal power plants—release air pollutants such as sulfur dioxide (SO_2) and nitrogen oxides (NO_x) high into the atmosphere to help limit the impact of these emissions on local air quality. Tall stacks can also increase the distance these pollutants travel in the atmosphere and harm air quality and the environment in downwind communities. The 1977 amendments to the Clean Air Act encourage the use of pollution control equipment over dispersion techniques, such as tall stacks, to meet national air standards. Section 123 of the Act does not limit stack height, but prohibits sources of emissions from using the dispersion effects of stack heights in excess of a stack's good engineering practice (GEP) height to meet emissions limitations.

GAO was asked to report on (1) the number and location of tall stacks of 500 feet or higher at coal power plants and when they began operating; (2) what is known about such stacks' contribution to the interstate transport of air pollution and the pollution controls installed at plants with these stacks; and (3) the number of stacks that were built above GEP height since 1988 and the reasons for this. GAO analyzed Energy Information Administration (EIA) data on power plants, surveyed states with tall stacks, and interviewed experts on the transport of air pollution. GAO is not making recommendations in this report. The Environmental Protection Agency and the Department of Energy stated they had no comments on this report.

Chapter 4 - Following its review of more than 2,000 scientific studies, on October 17, 2006, the Environmental Protection Agency (EPA) published its final revisions to the National Ambient Air Quality Standards (NAAQS) for particulate matter (particulates, or PM). Several states and industry, agriculture, business, and environmental and public health advocacy groups petitioned the US Court of Appeals for the District of Columbia Circuit, challenging certain aspects of EPA's revisions. A February 24, 2009, decision by the D.C. Circuit granted the petitions in part, denying other challenges, and remanded the standards to EPA for further consideration. While the court did not specifically vacate the 2006 PM standards, EPA initiated its next round of the periodic review of the particulates NAAQS, in part, in response to the court's decision, and plans to propose its decision regarding changes to the PM standards sometime during the fall of 2011. These actions, and the ongoing implementation of the 2006 PM NAAQS, have prompted renewed interest among Members of Congress.

Experiences and issues leading up to and following the promulgation of the 2006 PM2.5 NAAQS could provide relevant insights as EPA proceeds with its current review. Although a tightening of the standards, the 2006 particulates NAAQS are not as stringent as recommended by EPA staff or the independent scientific advisory committee mandated under the Clean Air Act (Clean Air Scientific Advisory Committee, or CASAC). The divergence from the CASAC's recommendations proved controversial, as did several other elements of the 2006 particulates NAAQS, including the decision not to exclude rural sources from the coarse particle standard.

EPA found that the evidence continued to support associations between exposure to particulates in ambient air and numerous health problems. Based on several analytical approaches, EPA estimated that compliance with the revised NAAQS would prevent 1,200 to 13,000 premature deaths annually, as well as substantial numbers of hospital admissions and missed work days due to illness. EPA revised the PM NAAQS by strengthening the 1997 standard for "fine" particulate matter 2.5 micrometers or less in diameter (PM2.5). Specifically, the agency lowered the allowable daily concentration averaged over 24-hour periods of PM2.5 in the air from 65 micrograms per cubic meter (μg/m3) to 35 (μg/m3). The annual PM2.5 standard, which is set in addition to the daily standard to address human health effects from chronic exposures to the pollutants, was unchanged from the 1997 standard. The decision not to tighten the annual standard was overturned by the D.C. Circuit and remanded to EPA for consideration.

The 2006 particulates NAAQS also retained the 24-hour standard and revoked the annual standard for slightly larger, but still inhalable, particles less than or equal to 10 micrometers (PM10). EPA abandoned its proposal to replace the particle size indicator of PM10 with a range of 10 to 2.5 micrometers (PM10-2.5). The D.C. Circuit's February 24, 2009, decision upheld EPA's decisions with regard to PM10 NAAQS.

EPA's ongoing implementation of the 2006 NAAQS, including EPA's November 13, 2009, final designation of those geographical areas not in compliance (typically defined by counties or portions of counties), has been an area of debate among some Members of Congress, states, and other stakeholders. Although EPA did not require new nonattainment designations for PM10, the tightening of the PM2.5 standard resulted in an increased number of areas in nonattainment compared to the designations for the 1997 PM NAAQS. EPA's November 2009 final designations for the 2006 PM NAAQS included 120 counties and portions of counties in 18 states as nonattainment areas based on 2006 through 2008 air quality monitoring data.

Chapter 5 - This chapter summarizes the Clean Air Act and its major regulatory requirements.

The principal statute addressing air quality concerns, the Clean Air Act was first enacted in 1955, with major revisions in 1970, 1977, and 1990. The Act requires EPA to set health-based standards for ambient air quality, sets deadlines for the achievement of those standards by state and local governments, and requires EPA to set national emission standards for large or ubiquitous sources of air pollution, including motor vehicles, power plants, and other industrial sources. In addition, the Act mandates emission controls for sources of 187 hazardous air pollutants, establishes a cap-and-trade program to limit acid rain, requires the prevention of significant deterioration of air quality in areas with clean air, requires a program to restore visibility impaired by regional haze in national parks and wilderness areas, and implements the Montreal Protocol to phase out most ozone-depleting chemicals.

This chapter describes the Act's major provisions and provides tables listing all major amendments, with the year of enactment and Public Law number, and cross-referencing sections of the Act with the major U.S. Code sections of the codified statute.

In: Federal Control of Air Emissions ISBN: 978-1-62081-808-4
Editors: H.M. Graham and T.E. Lee

Chapter 1

CONTROLLING AIR EMISSIONS FROM OUTER CONTINENTAL SHELF SOURCES: A COMPARISON OF TWO PROGRAMS — EPA AND DOI*

Jonathan L. Ramseur

SUMMARY

Air emissions from outer continental shelf (OCS) operations are subject to different regulatory programs, depending on the location of the operation.

The Department of the Interior (DOI) has jurisdiction over OCS sources in federal waters in the western Gulf of Mexico and most of the central Gulf. In addition, the Consolidated Appropriations Act, 2012 (P.L. 112-74), transferred air emission authority in the OCS off Alaska's north coast from the Environmental Protection Agency (EPA) to DOI. EPA has jurisdiction over sources in all other federal waters.

The primary difference between the EPA and DOI programs is rooted in the different statutory authorities: the 1990 Clean Air Act (CAA) and the 1978 Outer Continental Shelf Lands Act (OCSLA). The primary objectives of these statutes are different—air quality versus

* This is an edited, reformatted and augmented version of a Congressional Research Service publication, CRS Report for Congress R42123, dated January 9, 2012.

offshore energy development. The two regulatory programs reflect these underlying differences.

For much of the past 30 years, these differences received little attention, primarily because most of the federal oil and gas resources in EPA's jurisdiction have been subject to moratoria. In 2008, moratoria provisions expired, potentially opening many of the areas in EPA's jurisdiction to oil and gas leasing activity. If more OCS areas in EPA's jurisdiction are open for oil and gas leasing, policymakers interest in these differences will likely increase.

For OCS sources in EPA's jurisdiction, requirements depend on whether the source is located within 25 miles of a state's seaward boundary ("inner OCS sources") or beyond ("outer OCS sources"). Inner OCS sources are subject to the same requirements as comparable onshore emission sources, which vary by state and depend on the area's air quality status; outer sources are subject to various CAA provisions, including the Prevention of Significant Deterioration (PSD) program. In contrast, OCS sources in DOI's jurisdiction are subject to air emission requirements only if emissions would "significantly affect" *onshore* air quality.

A key difference between the EPA and DOI programs is the federal emission threshold that would subject a source to substantive requirements.

For sources in EPA's jurisdiction, this is the PSD threshold of 250 tons per year (tpy) of regulated emissions. Sources that exceed this level would likely be subject to Best Achievable Control Technology (BACT) and other provisions. States' analogous thresholds that apply to inner OCS sources may be more stringent. By comparison, a DOI OCS source applies an exemption formula, based on distance from shore (e.g., a source 30 miles from shore would have an emission threshold of 990 tpy).

If a source remains subject after this step, it must conduct air modeling to assess whether its emissions would have a significant affect on onshore air quality. In effect, this two-step process constitutes a much less stringent threshold than EPA's 250 tpy threshold.

Another substantial difference is the time frame allotted to the agencies for reviewing a potential source's permit (EPA) or activity-specific plan (DOI). In addition, the EPA permit process allows greater opportunity for input from the public. In particular, EPA's Environmental Appeals Board offers parties a powerful tool to compel agency review.

Therefore, two identical operations, located in separate jurisdictions, could face considerably different requirements and procedural time frames. Some stakeholders would likely argue that the additional opportunities for public involvement in EPA's permit process help create a balance between resource development and environmental concerns. Others would likely contend these steps present unnecessary burdens and timing uncertainty in the process.

INTRODUCTION

The same group of air pollutant emissions from outer continental shelf (OCS) operations are subject to different regulatory programs, depending on the location of the operation. The Department of the Interior (DOI) has jurisdiction over OCS sources in federal waters in the western Gulf of Mexico and most of the central Gulf. The Consolidated Appropriations Act, 2012 (P.L. 112-74), transferred air emission authority in the OCS off Alaska's north coast from the Environmental Protection Agency (EPA) to DOI. EPA has jurisdiction over sources in all other federal waters.

Figure 1 illustrates the divided jurisdiction in the Gulf of Mexico. Although active leases are located in EPA's jurisdiction, none of the Gulf of Mexico platforms are in EPA's jurisdiction.

Congress established the programs through different statutes, and the two agencies implement the programs through separate regulations. Pursuant to the underlying statutes, the regulations have different scopes, emission thresholds for eligibility, and compliance obligations.

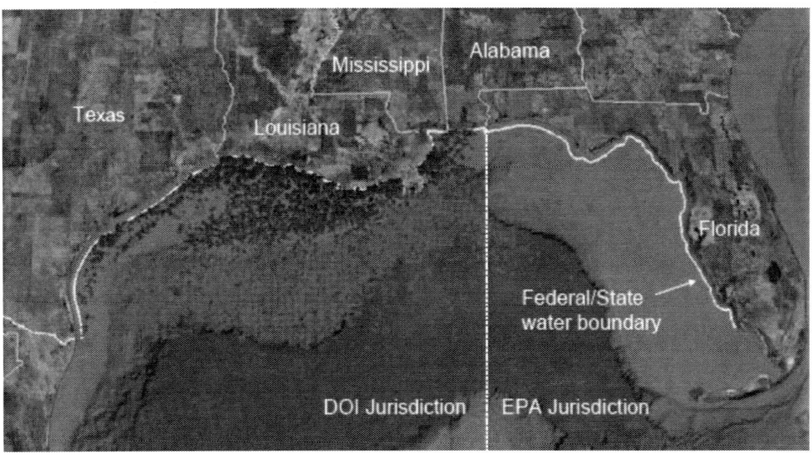

Source: Prepared by CRS; Active lease, platforms, and federal/state water boundary created Google Earth, using BOEMRE GIS files, at http://www.gomr.boemre. gov /homepg/data_center/geographic_mapping.html.

Notes: DOI has jurisdiction of OCS sources west of 87.5° longitude; EPA has jurisdiction east of this line. CRS inserted this line manually (in PowerPoint), so its position may not be precise.

Figure 1. EPA and DOI OCS Air Emission Jurisdictions in the Gulf of Mexico (red markers = active leases; green markers = platforms).

Therefore, two identical operations, located in separate jurisdictions, could face considerably different requirements and procedural time frames. Some may criticize this arrangement for its inconsistent treatment of air emissions. Others may point out that the differences create inconsistent opportunities for oil and gas development.

For much of the past 30 years, these differences received little attention, primarily because most of the federal oil and gas resources in EPA's jurisdiction have been subject to moratoria. In 2008, moratoria provisions expired, allowing many of the areas in EPA's jurisdiction to potentially open for oil and gas leasing activity.

According to DOI data from 2010, there were 3,409 *production* wells in the Gulf of Mexico and 23 in the Pacific.[1] Exploration drilling activity fluctuates: on December 16, 2011, 26 exploration wells were actively drilling in the Gulf.[2] All of these were in DOI's jurisdiction.

Estimates of oil and gas resources have spurred interest in areas other than the Gulf (*Table 1*). For example, OCS areas off the Atlantic coast have also received attention in recent years. In 2007, the Minerals Management Service (MMS) proposed a lease sale in an area offshore of Virginia.[3] The lease sale was set to take place in 2011.[4] In the aftermath of the 2010 *Deepwater Horizon* oil spill, the Obama Administration canceled the sale (May 27, 2010).[5]

Many Members of the 112th Congress have sought to expand OCS oil and gas development. The House has passed several bills (H.R. 1230 and H.R. 1231) that would direct DOI to hold lease sales in particular areas, including the formerly proposed Virginia lease sale and areas of Southern California.

In addition, areas of the U.S. Arctic region, which were not subject to the moratoria, have generated recent congressional interest as industry has sought to establish a greater presence in the area (e.g., H.R. 2021 and H.R. 2055). A primary driver for the activity in this region is the shrinking Arctic ice cap, or conversely, the growing amount of ice-free ocean in the summertime.[6]

As interest in developing OCS resources in EPA's jurisdiction has increased, the EPA OCS air program has received increased scrutiny from some Members. In particular, Members have focused on the permit activity of Shell in the Chukchi Sea, part of the federal waters off Alaska's north coast. In June 2011, the House passed H.R. 2021, which would amend EPA's OCS air emissions program. As mentioned above, H.R. 2055 would eliminate EPA authority in this region.

This report will provide an overview and brief history of the separate regulatory programs, examine and compare the requirements of the programs, and highlight potential implications of the differences. Although OCS sources can involve other industry sectors (e.g., deepwater ports or offshore wind projects), the focus of this report is air emissions from the OCS oil and gas exploration, development, and production operations.

Table 1. Estimates of Oil and Gas Resources in U.S. OCS Regions (2006 estimate)

OCS Region	Undiscovered Economically Recoverable Resources (at selected range of prices)	
	Oil (billion barrels)	Natural Gas (trillion cubic feet)
Alaska	8-22	27-94
Atlantic	2-3	14-21
Gulf of Mexico	36-40	163-202
Pacific	8-9	12-14

Source: Prepared by CRS from Minerals Management Service, Planning Area Resources Addendum to Assessment of Undiscovered Technically Recoverable Oil and Gas Resources of the Nation's Outer Continental Shelf, 2006.

Notes: The oil estimate is based on a price range of $46/barrel to $80/barrel. The natural gas estimate is based on a price range of $6.96/million cubic feet (Mcf) to $12.10/Mcf. In recent years, prices have gone both higher and lower than these ranges (See Energy Information Administration data at http://www.eia.gov).

Reader's Note: DOI Delegation

DOI created the Minerals Management Service (MMS) in 1982.[7] On May 19, 2010, during the Deepwater Horizon oil spill response, the Secretary of DOI replaced MMS with the Bureau of Ocean Energy Management, Regulation and Enforcement (BOEMRE). On October 1, 2011, DOI divided BOEMRE into three separate entities: the Bureau of Ocean Energy Management (BOEM), the Bureau of Safety and Environmental Enforcement (BSEE), and Office of Natural Resources Revenue (ONRR). Both BOEM and BSEE have responsibilities that related to air emissions from OCS sources. A DOI October 2011 direct final rule explains how existing OCS regulations are now divided (or shared in some cases) between the two new agencies.[8] Although MMS and BOEMRE no longer exist, this report refers to activities performed by these agencies during their existence. In addition, this report cites documents and websites that carry their names and are still valid.

OVERVIEW

OCS sources of air emissions can vary considerably, depending on the specifics of the operation. Offshore oil and gas sector operations, in particular, may include evolving technologies and take place in different settings, making it difficult to generalize air emission potentials. For example, mobile offshore drilling units involved in deepwater drilling in the Gulf of Mexico may emit considerably more emissions than a rig anchored to the sea floor in shallow water.

Operations in the Arctic region may involve icebreaker vessels, which may generate considerably more emissions than other support vessels.

A Senate Report from the 1990 Clean Air Act (CAA) amendments stated: "The construction and operation of OCS facilities emit a significant amount of air pollution which adversely impacts coastal air quality in the United States. Operational emissions from an OCS platform and associated marine vessels can routinely exceed 500 tons of nitrogen (NO_x) and one hundred tons of reactive hydrocarbons annually."[9]

Based on a review of EPA OCS permit documents,[10] the 1990 Senate Report estimate is still valid, particularly for NO_x emissions. *Table 2* identifies emission ranges of selected air pollutants from seven OCS oil/gas operations seeking, or recently receiving, air permits in EPA's jurisdiction. In some cases, the ranges cover a wide spectrum.

For each source, NO_x emissions account for the majority of the operations' regulated pollutants.

Four of the pollutants in *Table 2* are considered "criteria" pollutants and have a National Ambient Air Quality Standard (NAAQS).[11] Both NO_x and volatile organic carbon (VOC) emissions lead to ozone (O3) formation (sometimes described as smog), which is also a criteria pollutant.

According to congressional testimony from EPA Administrator Jackson, "a single exploratory drilling operation could emit approximately as much air pollution on a daily basis as a large state-of-the-art oil refinery."[12]

History of OCS Air Emission Governance

A brief history of federal OCS air emission governance may be instructive to policymakers. First, its history may help explain the rationale for the current framework. Second, some of the arguments made during its development may apply to current issues.

Table 2. Emission Ranges of Selected Air Pollutants from Outer Continental Shelf Oil/Gas Drilling Operations in EPA's Jurisdiction (emissions in tons per year)

Carbon Monoxide (CO)	Nitrogen Oxides (NOx)	Particulate Matter (PM_{10})	Sulfur Dioxide (SO2)	Volatile Organic Compound (VOC)
60-855	443-2339	6-57	1-833	11-96

Source: Prepared by CRS; potentials to emit (PTE) from EPA Regions 4 and 10 permit documents, particularly the "Statement of Basis" documents, provided with each draft permit. Data supplying the above ranges come from seven OCS oil/gas drilling permits. See permit materials from EPA Regions 4 and 10, at http://www.epa.gov/ region4/air/permits/OCSPermits/OCSpermits.html and http://yosemite.epa.gov/R10/AIRPAGE.NSF/Permits/ ocsap/. Accessed in September 2011.

Notes: The ranges are based on the operations' different potential to emit (PTE) values identified in the permit documents. Per 40 C.F.R. §52.21, PTE "means the maximum capacity of a stationary source to emit a pollutant under its physical and operational design. Any physical or operational limitation on the capacity of the source to emit a pollutant, including air pollution control equipment and restrictions on hours of operation or on the type or amount of material combusted, stored, or processed, shall be treated as part of its design if the limitation or the effect it would have on emissions is federally enforceable."

In several cases, EPA employed an air permit to require reductions from a pre-permitted PTE level, in order to avoid certain requirements (namely the PSD program, discussed below). The above ranges include the pre-permitted PTE values.

1953–1978

Passed as separate statutes in 1953, the Submerged Lands Act[13] and the Outer Continental Shelf Lands Act (OCSLA)[14] established federal jurisdiction and governance, respectively, for oil and gas resources on the OCS. The 1953 OCSLA did not specifically mention air emissions, but provided the Secretary of the Department of the Interior with general authority, stating: "The Secretary may at any time prescribe and amend such rules and regulations as he determines to be necessary and proper in order to provide for the prevention of waste and conservation of the natural resources of the outer Continental Shelf...."[15]

The Nixon Administration established the Environmental Protection Agency (EPA) in 1970 under an executive branch reorganization plan.[16] Although federal legislation addressing air pollution was first passed in 1955, the CAA Amendments of 1970 established the foundation of the federal air emissions program that exists today.[17]

For example, the 1970 law (among other provisions) set national standards for air quality, created a program to require the best available control technology at major new sources of air pollution, and established a program to regulate air toxics.

In 1977, Congress made further amendments to the CAA, establishing (among other provisions) the prevention of significant deterioration (PSD) program.

Tension arose between the CAA provisions of the 1970s and the 1953 OCSLA. The 1953 OCSLA did not specifically mention which federal agency would have jurisdiction over the air quality effects of OCS development, resulting in uncertainty over what jurisdiction, if any, the newly created Environmental Protection Agency (EPA) had for air quality control in the OCS.[18]

The jurisdictional uncertainty led EPA to assert its authority to regulate air emissions from OCS sources in April 1978.[19]

1978–1990

In September 1978, Congress passed the Outer Continental Shelf Lands Act of 1978 (P.L. 95- 372),[20] providing the Department of the Interior (DOI) with authority to regulate the national offshore oil and gas program including the regulation of air emissions from offshore oil and gas operations.[21] A federal appeals court affirmed DOI's authority (versus EPA authority) in 1979, in a case brought by the State of California against DOI.[22] DOI issued regulations in 1980.

Some stakeholders in California continued to voice concerns regarding the level of control for OCS source air emissions. A 1992 report from the National Research Council stated:

> Offshore air pollution is a problem in California that is not present in other oil-producing regions in the country (with the possible exception of the West Coast of Florida). Atmospheric conditions peculiar to coastal California involve a combination of the amount of sunlight, the mixture of pollutants already present, the direction of prevailing winds, and the presence of coastal mountain ranges—all of which work to trap smog along the coastal strip. A major problem here has been that air quality standards for air above federal waters were not as stringent as those for state waters.[23]

1990–Present

Congress made substantial amendments to the CAA in 1990.[24] A newly added Section 328 authorized EPA to implement certain air quality provisions of the act at offshore facilities, except for those west of 87.5 degrees longitude (i.e., Western and Central areas of the Gulf of Mexico). Offshore facilities not under EPA jurisdiction remain under DOI jurisdiction. EPA promulgated regulations for OCS sources in 1992.

In addition, Section 328(b) required the Secretary of the Interior to prepare a study assessing impacts of OCS sources in nonattainment areas that fail to meet NAAQS for ozone or nitrogen dioxide. Based on the results, the Secretary shall consult with the EPA Administrator to determine if additional actions are necessary. MMS published this study in 1995, which concluded that "the contribution of [OCS petroleum development] emission sources on onshore ozone concentrations is small."[25] The MMS regulations remained largely unchanged (discussed below).

Role of OCS Moratoria

Decades long moratoria of oil and gas development for much of the OCS have influenced the level of interest regarding OCS air emission issues. As a result of public laws and executive orders of the President, OCS moratoria along the Atlantic and Pacific coasts, parts of Alaska, and the Gulf of Mexico have been in place since 1982. On July 14, 2008, President Bush lifted the executive moratoria, which included planning areas along the Atlantic and Pacific coasts. On September 30, 2008, moratoria provisions in annual appropriations laws expired, allowing these areas to potentially open for oil and gas leasing activity.[31]

Thus, most of the OCS in EPA's jurisdiction has been off limits to oil and gas development until recently. Moreover, once a moratorium is removed, multiple legal and administrative processes must be satisfied before actual drilling can occur in a particular area.[32]

EPA OCS Air Program

CAA Section 328 provides the underlying authority for EPA's OCS program. Pursuant to this section, EPA established two regulatory regimes: one for OCS sources located within 25 miles of a state's seaward boundary ("inner OCS sources");[33] another for OCS sources located beyond 25 miles of a state's water boundary and extending to the boundary of the EEZ ("outer OCS sources").

EPA created the separate regimes based on the text of Section 328. This section directs EPA to develop regulations requiring all OCS sources "to attain and maintain Federal and State ambient air quality standards and to comply with the provisions of [the prevention of significant deterioration (PSD) program]."[34] In addition, Section 328 creates a more comprehensive program for inner sources, stating: "*Such requirements shall be the same as would be applicable if the source were located in the corresponding onshore area,* and shall include, but not be limited to, State and local requirements for emission controls, emission limitations, offsets, permitting, monitoring, testing, and reporting [emphasis added]."

EPA promulgated regulations—40 C.F.R. Part 55—in 1992 for OCS source air emissions.[35] Selected elements of these regulations are discussed below. In several instances, Part 55 directs OCS sources to air emission regulations in other CAA regulations, codified in separate parts of the *Code of Federal Regulations*. In general, these cross-references include the most substantive requirements for OCS sources.

Federal and State OCS Jurisdiction

The United States shares jurisdiction over its coastal waters with the coastal states. The 1953 Submerged Lands Act (SLA)[26] gave coastal states jurisdiction over the submerged lands, waters, and natural resources (e.g., oil deposits) located, in most cases, within 3 nautical miles off the coastline.[27] Within their offshore boundaries, coastal states have "(1) title to and ownership of the lands beneath navigable waters within the boundaries of the respective states, and (2) the right and power to manage, administer, lease, develop and use the said lands and natural resources...."[28] Accordingly, coastal states have the option of developing offshore oil and gas within their waters; if they choose to develop, they may regulate that development.

The waters, seabed, and natural resources beyond the states' waters are exclusively federal, and extend to the edge of the exclusive economic zone (200 nautical miles from shore).[29] However, the federal government maintains the authority to regulate commerce, navigation, national defense, power production, and international affairs within state waters.[30]

Outer OCS Source Requirements

Compared to inner OCS source requirements, outer OCS source requirements are fewer and *relatively* less complex, because sources need only comply with federal regulations. EPA argued in its rulemaking proposal that CAA Section 328 gave the agency "some latitude" to determine the regulations for outer sources.[36]

In addition to PSD requirements (which Section 328 specifically identified), EPA chose to apply New Source Performance Standards (NSPS), National Emission Standards for Hazardous Air Pollutants (NESHAPS), and CAA Title V permitting provisions to outer OCS sources. These are discussed below.

Prevention of Significant Deterioration

Pursuant to CAA Section 328 (and 40 C.F.R. Part 55) OCS sources are potentially subject to the CAA's Prevention of Significant Deterioration (PSD) program. Congress added the PSD program to the CAA in 1977[37] to address new or modified emission sources that would affect areas meeting National Ambient Air Quality Standards (NAAQS).[38] The primary objective of the program is to ensure that the air quality does not degrade in these areas with the addition of a new (or modified) source.

An OCS source that qualifies as a "major stationary source"[39] must comply with PSD provisions. The primary determinant is an annual emissions threshold.

For any stationary source, the threshold is 250 tons per year (tpy) of a regulated pollutant.[40] Some specific emission sources have a lower threshold of 100 tpy. Oil and gas exploration, development, and production activities are not among these specific sources.[41] Regardless, many OCS sources from the oil/gas industry are likely to approach or breach the 250 tpy threshold (*Table 2*).

In addition to the 250 tpy threshold, as of January 2, 2011, an OCS source must consider its greenhouse gas (GHG) emissions.[42] These emissions are subject to a different threshold. Per a 2010 rulemaking, EPA is phasing in GHG applicability for the PSD program.[43] As of July 1, 2011 (the start of the second of four phases), new emission sources—not already subject to PSD for other pollutants—would be subject to PSD, if GHG emissions equal or exceed 100,000 tpy of carbon dioxide equivalents (CO2e).[44] If an OCS source is already subject to PSD for one or more of the 250 tpy-threshold pollutants, the GHG emission threshold is 75,000 tpy of CO2e.[45]

If an OCS source meets the definition of a "major stationary source," several PSD requirements apply. Selected requirements are discussed below, followed by an exemption for certain OCS sources.

Best Available Control Technology (BACT)

A source must apply BACT for each regulated pollutant that would be emitted in significant amounts.[46] This means that a source may qualify as a "major stationary source" based on only one pollutant, but may need BACT for multiple pollutants.[47] The significant amounts for selected pollutants are listed in *Table 3.*

Table 3. Selected Significant Emission Rates in PSD Program (tons per year)

CO	NOx	SO2	PM	PM10	PM2.5	Ozone
100	40	40	25	15	10[a]	40[b]

Source: 40 C.F.R. §52.21(b)(23).

[a] 10 tpy of direct $PM_{2.5}$ emissions; 40 tpy of SO_2 emissions; 40 tpy of NO_x emissions unless demonstrated not to be a PM2.5 precursor.

[b] 40 tpy of VOC compounds or NO_x.

BACT is determined on a case-by-case basis, taking into account a proposed control measure's energy, environmental, and economic impacts.[48] Most states have the authority to implement the PSD program and set BACT for applicable sources, but as mentioned above, only a handful of locales have Part 55 (OCS regulatory) authority. Therefore, the applicable EPA Region would determine BACT for OCS sources in most locations.[49] Due to the moratoria, EPA has made such determinations in only a few instances.

Air Quality Demonstration and Analyses

A source must demonstrate that its emissions would not cause or contribute significantly to a violation of the NAAQS[50] or any allowable maximum increase over the baseline concentration.[51] To meet this demonstration, a source must provide an air quality analysis for each pollutant it would emit in a significant amount.[52] In general, the pollutant analysis must include at least one year of continuous air quality monitoring data.[53]

A source must provide further analysis regarding impacts to visibility and soils and vegetation.[54] If a source's emissions may impact a Class I area—156 listed national memorials, monuments, parks, and wilderness over certain sizes—[55] EPA must consult with the applicable Federal Land Manager. The Manager has the opportunity to demonstrate that the source would yield adverse impacts, such as reduced visibility. If EPA accepts this demonstration, the source would not receive a permit. If EPA disagrees, the agency must provide a publicly available explanation.[56]

Emissions Monitoring and Reporting

Pursuant to PSD regulatory authority, EPA may require a source to continue to monitor ambient air emissions during its operation to determine the effect of the emissions.[57] EPA may exempt a source from this requirement, if a source's emission levels are below pollutant-specific thresholds.[58] Part 55 also contains provisions authorizing EPA to require emissions monitoring.[59]

Exemption for Temporary Sources

OCS exploratory drilling operations may qualify for an exemption from certain PSD program requirements. The exemption has two conditions. First, regulated emissions from the major stationary source must not impact a Class I area (discussed above). Second, the source's emissions must be "temporary."[60] EPA has not defined "temporary" in the PSD regulations, but in a 1980 *Federal Register* preamble, EPA stated that it considered sources operating for less than two years in a given location to be temporary sources.[61]

Sources meeting the conditions of the exemption are not subject to the air quality demonstration and analyses discussed above. Several of OCS exploratory drilling operations that received EPA air permits qualified as temporary sources.

New Source Performance Standards

Part 55 requires OCS sources to comply with applicable New Source Performance Standards (NSPS) in 40 C.F.R. Part 60. NSPS are nationally uniform, technology-based standards for specific categories of stationary emission sources.[62]

The format of the standard can vary from source to source, and may entail a numerical emission limit, a design standard, an equipment standard, or a work practice standard.

Offshore oil development operations may involve equipment that has a NSPS.

For example, several OCS draft permits prepared by EPA identify specific internal combustion engines with a NSPS in 40 C.F.R. Part 60, Subpart IIII.[63]

National Emission Standards for Hazardous Air Pollutants

OCS sources may be subject to National Emission Standards for Hazardous Air Pollutants (NESHAP). Part 55 requires compliance with any regulations promulgated pursuant to CAA Section 112,[64] if they are "rationally related to the attainment and maintenance of Federal or State ambient air quality standards or the requirements of [the PSD program]."[65]

EPA has promulgated multiple regulations pursuant to CAA Section 112 (e.g., 40 C.F.R. Parts 61 and 63). In general, these regulations establish emissions standards for hazardous air pollutants released from specific sources (e.g., industries or equipment). Whether these regulations apply depends on the operating equipment of the specific OCS source.

Based on OCS permit documents from EPA Region 4 and Region 10, existing provisions in Part 61 are unlikely to apply to an OCS oil/gas drilling operation.[66] Several of the permits identify equipment that would be potentially subject to provisions in Part 63 (e.g., Subpart ZZZZ, which regulates reciprocating internal combustion engines). However, in many cases the units satisfy Subpart ZZZZ requirements by complying with the NSPS in 40 C.F.R. Part 60, Subpart IIII (mentioned above).

Title V/Part 71 Operating Permits

A source may require a Title V permit, but not be subject to the provisions of other parts of the CAA, such as BACT under the PSD program. The emissions threshold for a regulated pollutant is lower for Title V than the PSD program. Instead of 250 tpy (as discussed above), the Title V threshold is generally 100 tpy.[67]

Part 55 directs outer OCS sources to 40 C.F.R. Part 71, which contains the regulations promulgated to CAA Title V. The primary purpose of the Title V permit program is administrative/enforcement efficiency. Title V permits collect all applicable requirements in a single permit. Title V permits are not intended to change or alter the existing, underlying requirements or add requirements.[68]

Although some may describe the Title V permit provisions as non-substantive, others may view them as onerous. Title V permits are issued for five-year periods and must be renewed thereafter. Permits generally require the following:

- emissions limitations and standards to assure compliance with all applicable requirements;
- emissions monitoring, recordkeeping, and reporting;

- fee payments to support administrative expenses; and
- annual certification by a responsible official of the source.

Inner OCS Source Requirements

Inner OCS sources are subject to all of the requirements for outer OCS sources (described above) and any applicable state and/or local air emissions requirements. In the event of conflict between federal and state/local regulations, the more stringent provisions would apply.[69]

Attainment vs. Nonattainment Areas

Compared to outer OCS sources, the location of the inner OCS source plays a key role in determining its applicable emission requirements. This is a function of the CAA and the potential variety of regulations among states and locales.

A fundamental element of the CAA, which is generally implemented by states (at least for onshore sources), is its approach of setting more stringent standards for areas not meeting certain air quality standards ("nonattainment areas").[70] For example, in a nonattainment area, an inner OCS source may be subject to the lowest achievable emission rate (LAER), which is by definition more stringent than BACT.[71] In addition, emissions from new or modified sources must also be offset by reductions in emissions from existing sources.

For the OCS air permits recently issued by EPA—in the eastern Gulf of Mexico and off the northern Coast of Alaska (Beaufort and Chukchi Seas)—the nearby coastal areas are in attainment.[72] In general, the coastal non-attainment areas for pollutants common to OCS oil/gas operations (e.g., ozone) are located in California and the Mid Atlantic (*Figure 2*).[73] If OCS operations expand, nonattainment areas may become more of a factor.[74]

Notice of Intent/Corresponding Onshore Area Designation

An owner/operator of an inner OCS source must submit a Notice of Intent (NOI) to the appropriate EPA regional office and the appropriate state agency (or agencies) of the nearest onshore area and adjacent areas. Among other elements, the NOI must include (1) an estimate of the source's potential emissions (in tons per year) of any air pollutant; and (2) information allowing for an analysis of the source's impact in onshore areas.[75] As discussed below, the NOI must include emission information from vessels associated with the operation.

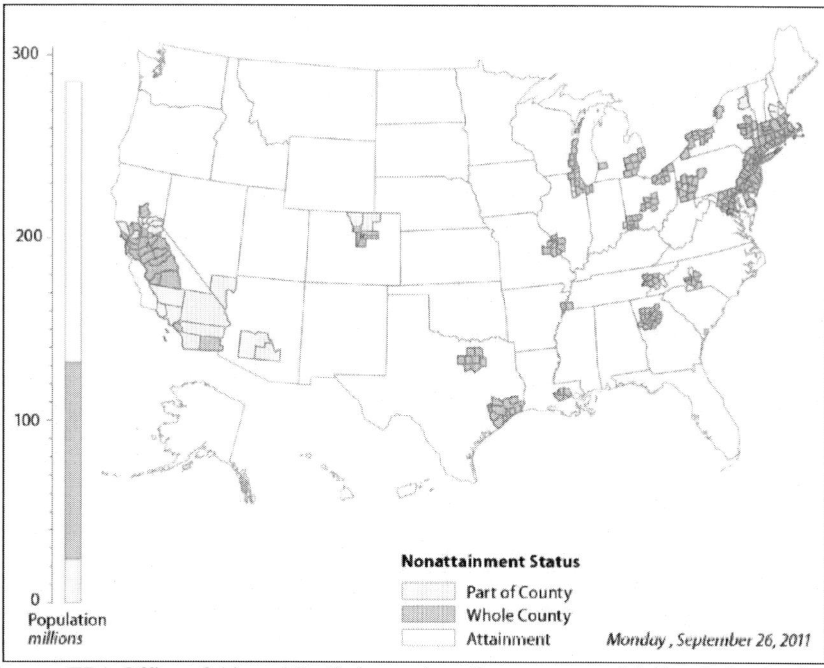

Source: EPA Office of Air and Radiation, at http://www.epa.gov/air/data/index.html.

Note: EPA's nonattainment map for ozone is based on its 1997 ozone NAAQ, which was lowered from 0.08 parts per million (ppm) to 0.075 ppm in 2008. See CRS Report R41062, Ozone Air Quality Standards: EPA's Proposed Revisions, by James E. McCarthy.

Figure 2. Map of Non-Attainment Areas for Ozone (8-Hour).

The main purpose of the NOI is to allow the applicable state agency to decide if it wants to submit a request as corresponding onshore area (COA).[76] The CAA requires OCS sources to comply with air emission provisions applicable to onshore sources in the COA. In general, the onshore area closest to the OCS source is considered the COA. The act provides EPA with the authority to designate another area as the COA, if that area has more stringent air emission controls and that area will be impacted by emissions from the OCS source.[77] For this to occur, the state agency from the state seeking a COA redesignation must make a formal request to EPA. EPA must make a COA designation determination—based on state agency documentation and following a public comment period—within 240 days of receiving the NOI from the OCS source.

EPA regulations do not allow a COA redesignation for exploratory OCS sources (e.g., exploratory drilling operations). The statute is silent on this issue, but EPA determined it was unreasonable to require an exploratory source, which (according to EPA) may operate for 3-4 months, to undergo an administrative procedure that can last for 8 months.[78]

OCS Vessel Emissions

Air emissions from vessels associated with OCS operations have generated some interest and debate.[79] In some offshore oil and gas operations, the emissions from vessels may represent a substantial percentage of the operation's overall air emissions. Section 328(a)(4)(C) states:

The terms "Outer Continental Shelf source" and "OCS source" include any equipment, activity, or facility which—

a) emits or has the potential to emit any air pollutant,
b) is regulated or authorized under the Outer Continental Shelf Lands Act [43 U.S.C. 1331 et seq.], and
c) is located on the Outer Continental Shelf or in or on waters above the Outer Continental Shelf. Such activities include, but are not limited to, platform and drill ship exploration, construction, development, production, processing, and transportation. For purposes of this subsection, emissions from any vessel servicing or associated with an OCS source, including emissions while at the OCS source or en route to or from the OCS source within 25 miles of the OCS source, shall be considered direct emissions from the OCS source.[80]

While crafting the regulations (40 C.F.R. Part 55) implementing CAA Section 328, EPA interpreted the above provision to include vessels as OCS sources only when (1) they are permanently or temporarily attached to the seabed for the purpose of exploring, producing, or developing natural resource activities; or (2) they are physically attached to an OCS facility.[81] If a vessel qualifies as an OCS source, it may need to comply with other sections of the CAA, namely the PSD program and BACT.

Emissions from support vessels (i.e., non-OCS source vessels) play a role in how the OCS source the vessels support is regulated. All air emissions, including transit-related emissions within 25 miles of the OCS source, from support vessels must be accounted for in an OCS source's "potential to emit."

In some situations, the support vessel emissions may cause the OCS source's emissions to cross regulatory thresholds (i.e., 250 tpy of regulated pollutants in the case of PSD).[82]

State Authority

Coastal states may seek authority to implement and enforce air emission requirements for OCS sources in federal waters adjacent to state waters.[83] This authority can include sources within the 25 mile seaward boundary and/or sources beyond this boundary.[84] To obtain this authority, a governor must submit his/her state regulations to the EPA Administrator, who determines if the state provisions are adequate, based on specific criteria.[85] If a state does not seek delegation, EPA implements the program in waters adjacent to that state. In addition, if a neighboring state has sought and received a COA redesignation (as discussed above), EPA may implement the more stringent standards of this neighboring state.[86]

According to a 2008 EPA *Federal Register* notice, only four local air pollution control agencies in California[87] have received authority to implement Part 55.[88] This relatively small number is likely related to the moratoria. EPA implements and enforces the regulations for all other sources within EPA's jurisdiction, including the state/local air emission regulations for inner OCS sources. If policymakers open more OCS areas to oil/gas activities, this dynamic may change.

Procedural Requirements

Although Part 55 does not include specific administrative procedures and public participation requirements, statutory provisions and requirements referenced in Part 55 would apply. *Table 4* identifies procedural time frames for PSD and Title V permits that would apply to outer OCS sources. Inner OCS sources would be subject to the same standards as onshore sources. These standards vary by state and may be more stringent than the time frames identified below.

Public Participation

In general, when EPA prepares a draft permit, the agency must provide 30 days for public comment.[89] Parties may request a public hearing,[90] which EPA must hold if the Administrator "finds, on the basis of requests, a significant degree of public interest in a draft permit."[91]

In addition, the agency has the discretion to extend the comment period beyond 30 days,[92] and it may reopen the comment period to expedite the decision process.[93]

Table 4. EPA Review of PSD and Title V Permits (outer OCS sources)

Time Frames	PSD Permit	Title V Permit[a]
For EPA to review permit to determine if application is complete	30 days[b]	Deemed complete within 60 days of receipt, unless EPA determines otherwise within that time frame[c]
After deemed submitted, public comment period	30 days (with extension at EPA's discretion)[d]	30 days (with extension at EPA's discretion)[e]
Final Decision	1 year after filing of completed permit application[f]	18 months after receiving a completed permit application[g]

Source: Prepared by CRS.

Note: Inner OCS sources would be subject to state/local air emission requirements, which may include more stringent time frames than identified above.

[a] OCS sources can apply for a Title V permit within 12 months after first becoming a source (40 C.F.R. §71.5).

[b] 40 C.F.R. §124.3.

[c] 40 C.F.R. §71.5

[d] 40 C.F.R. §124.10

[e] 40 C.F.R. §71.11.

[f] 42 U.S.C. §7475(c).

[g] 40 C.F.R. §71.7.

Administrative Appeals Process

PSD and Title V permits are subject to an administrative appeals process. Any person who submitted comments concerning the draft permit (or participated in a public hearing) may petition—within 30 days of the final permit decision—EPA's Environmental Appeals Board (EAB) to review any condition of the permit.[94]

In addition, the EAB may decide on its own initiative to review conditions in a permit. The EAB must issue its decision "within a reasonable time" after receiving a petition.[95]

Judicial Review

In addition, parties may contest EPA's permit decision in the court system. Judicial review of permit decisions is typically governed by the particular environmental statute that is the subject of the litigation.[96] CAA Section 307 provides for judicial review of EPA actions. In addition, CAA Section 502 includes judicial review provisions for Title V permit activity, providing an opportunity for judicial review in state court of final permit actions by the applicant, persons who participated in the public comment process, and any other person who could obtain judicial review of such actions under state laws.[97] However, parties must go through the EAB process before seeking judicial review of agency action.[98]

DOI OCS AIR PROGRAM

The DOI authority to address OCS air emissions comes from the 1978 OCSLA, which directs the DOI Secretary to promulgate regulations "for compliance with the national ambient air quality standards pursuant to the Clean Air Act (42 U.S.C. 7401 et seq.), to the extent that activities authorized under this subchapter significantly affect the air quality of any State."[99]

To satisfy this directive, DOI (acting through the U.S. Geological Survey) promulgated regulations in 1980.[100] The regulations were redesignated (i.e., renumbered) in 1998,[101] but the 1980 provisions generally remain the same in 2011. The recent administrative changes (described in the earlier "Reader's Note") have led to a further restructuring of the regulations. An October 2011 direct final rule separated BOEM and BSEE regulations, creating 30 C.F.R. Part 550 for BOEM provisions (BSEE regulations were retained in their original location).[102] The rule did not add or remove OCS requirements, but makes organizational changes needed to accommodate the DOI administrative changes.

In addition, DOI or its agencies (e.g., MMS, BOEM) have periodically issued Notices to Lessees and Operators (NTLs), which "clarify, supplement, or provide more detail about certain requirements."[103]

Activity-Specific Plan Requirements

Before conducting operations on the OCS, leaseholders must (among other requirements) submit and receive approval for activity-specific plans.

For example, the OCSLA requires lessees to have an approved Exploration Plan (EP) for exploration activities[104] and a Development and Production Plan (DPP) for development and production activities[105] or a Development Operations Coordination Document (DOCC) in areas, such as portions of the Western Gulf of Mexico, where significant activities have already taken place.[106] (Hereafter in this report, references to DPPs include DOCCs.) BOEM is charged with reviewing and approving these plans.[107]

These plans must include, among other provisions, a facility's projected emissions (in tpy and other measurements) of SO_2, particulate matter (measured in $PM_{2.5}$ and PM_{10} when applicable), NO_x, CO, and volatile organic compounds (VOC).[108] The lessee will apply these projections, and other related information (e.g., distance from shore), to determine whether certain requirements apply (e.g., BACT). A 2009 NTL stated that these requirements apply "regardless of whether the proposed activities are in an area under [DOI] jurisdiction or EPA air quality jurisdiction."[109]

Exemption Determination

A primary determination is whether the OCS source is exempt from further air emission requirements. The DOI regulations contain an exemption formula, based on projected emissions and distance from shore.[110] For all of the pollutants listed above except CO emissions, the exemption formula is:

$$E = 33.3D$$

where:

E equals the exemption threshold (in tpy), and
D equals distance from shore (in miles).

For example, if an OCS source would be located 30 miles from shore, it would be exempt from further air emission requirements as long as the projected emissions for each pollutant (SO_2, PM, NO_x, and VOC) were below its exemption threshold of 990 tpy.[111] Note the primary threshold for EPA's substantive requirements (e.g., PSD program) is 250 tpy, and states may have even lower thresholds that would apply to inner OCS sources.

Significance Determination

For a non-exempt OCS source, the next determination is whether projected air emissions from any pollutants would "significantly" affect onshore air quality (VOC emissions from non-exempt sources are automatically deemed to significantly affect onshore air quality). These sources must make the significance determination by using an approved air quality model.[112]

If the model indicates that air pollutant concentrations—specified in 40 C.F.R. §550.303(e)—would be exceeded, the facility's emissions would significantly affect onshore air quality. The DOI significance levels are the same required by EPA when determining whether a new major source would significantly impact the air quality of a neighboring nonattainment area.[113]

Required Emission Controls

If air emissions from a non-exempt OCS source would significantly affect the air quality of an onshore area (based on the modeling described above), further requirements apply. The requirements depend on whether the impacted onshore location is an attainment (or unclassifiable) or nonattainment area. If more than one area would be impacted, the more stringent requirements would apply.[114]

Emissions Monitoring and Reporting

According to the DOI regulations, lessees must monitor air emissions ("in a manner approved or prescribed by the Regional Supervisor") regardless of the source's exempt status or whether the OCS source's emissions would significantly impact air quality.[115]

In addition, the regulations require lessees to submit emission information to DOI on a monthly basis. However, it is uncertain whether OCS sources are complying with this provision as written.

Nonattainment Areas

For nonattainment areas (e.g., *Figure 2*), the source must "fully reduce" emissions of any pollutant that would significantly affect air quality. To achieve this objective, the lessee must apply BACT, which, like EPA's definition, allows for the consideration of energy, environmental, and economic impacts. The lessee determines BACT, but a regional BOEM official verifies BACT on a case-by-case basis.[116]

By comparison, EPA imposes LAER for new (inner OCS) sources in nonattainment areas, a potentially more stringent control. However, if BACT alone would not address the applicable emissions at an OCS source in DOI's jurisdiction, the source must make additional reductions or acquire emission offsets from another source.[117]

Attainment Areas

If attainment areas would be affected, an OCS source must apply BACT to emissions of any air pollutant that would significantly affect air quality in the attainment area. In addition, the lessee must determine whether—after applying BACT—emissions of total suspended particulates (TSP) or SO_2 would increase emission concentrations over specified thresholds ("maximum allowable increases") in the applicable area.[118] The specified maximum allowable increases are similar to those in EPA's PSD regulations.[119] However, the EPA regulations address PM_{10} instead of TSP emissions, and also include NO_2 emissions.

The likely explanation for this difference is that the DOI regulations reflect the EPA standards in place in 1980. Since that time, EPA has revised standards and added new pollutants.[120]

Temporary Facilities

OCS sources that are considered "temporary facilities" must apply BACT to address emissions of any pollutant that would significantly affect the air quality of an onshore area.[121] Temporary facilities are defined as offshore operations in one location lasting less than three years.[122] Note that EPA's time frame for temporary is two years.

OCS Vessel Emissions

OCS vessel emissions in DOI's jurisdiction are counted the same as emissions in EPA's jurisdiction. CAA Section 328(a)(4) contains the definition for OCS source. This definition applies to sources in both EPA and DOI jurisdictions.[123]

As discussed above, support vessel emissions "servicing or associated with an OCS source, including emissions while at the OCS source or en route to or from the OCS source within 25 miles of the OCS source, shall be considered direct emissions from the OCS source."

The DOI regulatory text addressing vessel emissions does not precisely match up with the CAA text. Although the activity-specific plan regulations require lessees to document support vessel emissions,[124] the exemption formula provision (discussed above) specifically points to air emissions documented under a different subsection,[125] which does not include vessel emissions. However, DOI *guidance* on emission calculations for plan submittals mirrors the statutory text.[126]

Procedural Requirements

The DOI regulations set specific time frames for agency review of Exploration Plans (EPs) and Development and Production Plans (DPPs). In general, these time frames are considerably shorter than those for EPA. These time frames are identified in *Table 5.* The maximum time frame for EP review is much shorter than for DPP review.

Table 5. BOEM Review of Exploration Plans and Development and Production Plans (time frames for different procedural steps)

Time Frames	Exploration Plan	Development and Production Plan
For BOEM to determine if a plan is deemed submitted	15 working days	25 working days
After deemed submitted, for a governor of an affected State to submit comments	21 calendar days	60 calendar days
After deemed submitted, public comment period	No public comment period	60 calendar days (concurrent with state governor time frame)
Final Decision[a]	30 calendar days after plan is deemed submitted	60 calendar days after (1) comment period closes; (2) BOEM releases its final environmental impact statement (EIS);[b] or (3) BOEM receives plan amendments

Source: 30 C.F.R. §§550.231-550.233 (EP) and §§550.266-550.270 (DPP).

[a] BOEM may (1) approve the plan; (2) require a plan modification; or (3) disapprove plan.

[b] The regulations do not require an EIS for every DPP. In general, DOI must prepare an EIS at least once in every OCS planning area (30 C.F.R. §550.269).

Coastal Zone Management Act Review

Section 307(c)(3) of the Coastal Zone Management Act (CZMA) requires lessees to certify the "consistency" of an EP or DPP with an approved coastal zone management program of an affected state (or states).[127] Section 307(f) specifically requires coordination with CAA requirements.

When a state determines that a lessee's plan is inconsistent with its coastal zone management program, the lessee must either reform its plan to accommodate those objections and resubmit it for BOEM and state approval, or succeed in appealing the state's determination to the Secretary of Commerce.[128] Some have questioned the impact of the CZMA consistency provision.[129] Historically, states have concurred with about 95% of the federal actions they have been asked to certify.

As of 2009, 43 consistency decisions have been subjects of Commerce Secretary determinations. The subject of 18 of these appeals has been offshore energy activities. However, the most recent of these decisions was rendered in 1999.[130] It is widely believed that the existence of the consistency requirement and the uncertainty of the outcome of an appeal have led applicants to negotiate with states and to modify proposed actions early on, thereby reducing the number of appeals. However, there are no data on the number of proposed actions that have been altered because of the consistency process.[131]

In addition, the CZMA regulations—implemented by the National Oceanic and Atmospheric Administration (NOAA)—indicate that some OCS activities may not require a review.[132] As activities occur farther offshore, a state's CZMA review authority becomes less clear. According to NOAA, a coastal state's ability to review activity-specific plans stops at the point where coastal effects are not "reasonably foreseeable." Whether this threshold is met would be determined on a case-by-case basis with the state, the lessee, and DOI.[133]

Public Participation

The DPP regulations require a 60-day comment period, in which any party may submit comments or recommendations to the BOEM Regional Supervisor.[134] The EP process does not include an opportunity for public participation (*Table 5*).

Judicial Review

The OCSLA provides opportunity for judicial review of agency action alleged to be in violation of federal law, including the OCSLA, its implementing regulations, and the terms of any permit or lease.[135]

Comparison of DOI and EPA Air Programs

Table 6 compares selected elements of the EPA and DOI air emission programs. The table examines the requirements applicable to outer OCS sources in EPA's jurisdiction. Inner OCS sources must follow the requirements that would apply to an onshore source in the corresponding onshore area. As discussed above, these provisions vary by state and whether the area is an attainment or nonattainment area. Regardless, inner OCS source requirements would be at least as stringent as outer source requirements in all situations, and potentially more stringent in some locations (e.g., California).

The primary difference between the EPA and DOI programs is rooted in the different statutory authorities: the 1990 CAA and the 1978 OCSLA. The primary objectives of these statutes are different—air quality versus offshore energy development. The two regulatory programs reflect these underlying differences.

OCS sources in EPA's jurisdiction must comply with the PSD program (e.g., BACT), among other requirements. Sources subject to DOI's regime must comply with similar standards (e.g., BACT), only if their emissions would "significantly affect" a state's air quality. This key difference translates into a considerably different scope of applicability between the two programs. For example, DOI's two-step significance determination is a potentially much less stringent threshold than EPA's 250 tpy threshold for its PSD program. Moreover, the federal threshold for Title V permits is 100 tpy. For sources within 25 miles of a state's shores, the PSD and Title V thresholds may be even lower.

Another substantial difference is the time frame allotted to the agencies for reviewing a potential source's permit (EPA) or activity-specific plan (DOI). In addition, the EPA permit process allows greater opportunity for input from the public. In particular, EPA's EAB offers parties a powerful tool to compel agency review. Some stakeholders would likely argue that the additional opportunities for public involvement in EPA's permit process help create a balance between resource development and environmental concerns. Others would likely contend these steps present unnecessary burdens and timing uncertainty in the process.

If more OCS areas in EPA's jurisdiction are open for oil and gas leasing, policymakers interest in these differences will likely increase.

Table 6. EPA and DOI OCS Air Emission Programs (comparison of selected elements)

Program Elements	EPA Outer OCS Sources Program[a]	DOI
Underlying statutory citation	1990 CAA §328 (42 U.S.C. §7627)	1978 OCSLA §5(a)(8) (43 U.S.C. §1334(a)(8))
Underlying statutory authority	Directs EPA to develop regulations requiring all OCS sources "to attain and maintain Federal and State ambient air quality standards and to comply with the provisions of [the PSD program]"	Directs DOI to develop regulations for compliance with CAA national ambient air quality standards, "to the extent that activities authorized under this subchapter significantly affect the air quality of any State"
Date of implementing regulations	September 4, 1992	March 7, 1980
Jurisdiction	All OCS sources in federal waters, except those west of 87.5 degrees longitude (the western and most of the central Gulf of Mexico) and in the federal OCS off Alaska's north coast[b]	All OCS sources in Gulf of Mexico federal waters that are west of 87.5 degrees longitude (the western and most of the central Gulf of Mexico) and OCS sources in federal waters off Alaska's north coast
Framework of requirements	Air emissions permit: PSD and/or Title V *OCS sources in EPA's jurisdiction must also submit applicable activity-specific plans per DOI regulations*[c]	Activity-specific plans: Exploration Plan (EP) or Development and Production Plan (DPP)
Emission thresholds for substantive requirements (e.g., BACT)	250 tpy	Two-step determination: 1) are emissions exempt based on distance from shore?[d] 2) if not exempt, would emissions "significantly" affect onshore air quality (as determined by modeling)?

Table 6. (Continued)

Program Elements	EPA Outer OCS Sources Program[a]	DOI
Emission monitoring and reporting	Required for Title V permits (100 tpy threshold) and PSD permits (250 tpy threshold)	Monitoring and monthly reporting required regardless of significance determination;[e] but it is uncertain whether this is occurring
Pollutants subject to thresholds	Per PSD regulations, any "regulated pollutant,"[f] including those with a national ambient air quality standard (NAAQS): • Sulfur dioxide (SO_2) • Particulate matter ($PM_{2.5}$ and PM_{10}) • Nitrogen dioxide (NO_2) • Carbon monoxide (CO) • Ozone • Lead • Any pollutant identified as a onstituent/precursor to the above (e.g., volatile organic compounds (VOC)) • Greenhouse gases (GHG)[g]	Per DOI regulations: • SO_2 • $PM_{2.5}$ and PM_{10} • Nitrogen oxide (NOx) • CO • VOC[h]
Substantive requirements if thresholds met	PSD permit requirements, including: 1) BACT 2) Air quality demonst ra-tion/analysis	If affected onshore area is an attainment area: BACT If affected onshore area is a nonattainment area: BACT and reduce all emissions with additional reductions or offsets If more than one area is impacted, the more stringent requirements would apply
Temporary source exemption	If operating for less than two years in a given location, sources are subject to BACT, but not the air quality demonstration and analyses	If operating in one location less than three years, a source must apply BACT to address emissions of any pollutant that would significantly affect the air

Program Elements	EPA Outer OCS Sources Program[a]	DOI
		quality of an onshore area
Other potentially applicable air emission requirements	New Source Performance Standards National Emission Standards for Hazardous Air Pollutants Title V permits Coastal Zone Management Act Review per DOI regulations	Activity-specific plans must include air emission information, documenting exemption Coastal Zone Management Act Review could potentially lead to air emission modifications
Time frames for agency review	PSD permit determination within 1 year of complete submittal; Title V permit within 18 months	Activity-specific plans have agency review deadlines (e.g., BOEM must provide an interim/final decision for an EP within 30 days of it being submitted)
Other federal agency involvement	EPA must consult with the applicable Federal Land Manager if a source's emissions may impact a Class I area	No analogous authority
State implementation	Coastal states may seek authority to implement and enforce EPA requirements for OCS sources in federal waters adjacent to state waters	No analogous authority
Opportunities for public participation	EPA agency must provide a 30-day public comment period when it issues a permit	BOEM must provide a 60-day public comment period for parties to review a DPP
Opportunities for administrative appeal	Environmental Appeals Board: any person can appeal an agency action	No analogous process
Opportunity for legal challenge	CAA provides opportunity for judicial review of agency actions	OCSLA provides opportunity for judicial review of agency actions

Source: Prepared by CRS.

[a] Pursuant to CAA Section 328, EPA established two regulatory regimes: one for OCS sources located within 25 miles of a state's seaward boundary ("inner OCS sources"); another for OCS sources located beyond 25 miles of a state's water boundary and extending to the boundary of the EEZ ("outer OCS sources"). Requirements for "inner sources" are the same as would be applicable if the source were located in the corresponding onshore area. These requirements will

vary by state and whether the corresponding onshore area is an attainment or nonattainment area for regulated pollutants.

[b] P.L. 112-74 (signed by President Obama December 23, 2011) transferred air emission authority from EPA to DOI for OCS sources off Alaska's north coast.

[c] The DOI activity-specific plans entail multiple provisions, including potential air emission requirements. See Notice to Lessees and Operators 2009-N11, "Air Quality Jurisdiction on the OCS," December 4, 2009.

[d] For all but CO emissions, the exemption formula is: E = 33.3D. Thus, a source located 30 miles from shore would be exempt if its emissions were above 990 tpy.

[e] This appears to apply regardless of the source's exempt status or whether the OCS source's emissions would significantly impact air quality (30 C.F.R. §550.303(k)).

[f] The official term is "regulated NSR [New Source Review] pollutant," which includes (among others) any pollutant for which a national ambient air quality standard (NAAQS) has been promulgated and any pollutant identified as a constituent or precursor for a regulated NSR pollutant (40 C.F.R. §52.21(b)(50)).

[g] As of January 2, 2011, an OCS source must consider its GHG emissions. These emissions are subject to a different threshold. As of July 1, 2011, new emission sources—not already subject to PSD for other pollutants—would be subject to PSD, if GHG emissions equal or exceed 100,000 tpy of carbon dioxide equivalents ($CO2_e$). If an OCS source is already subject to PSD for one or more of the 250 tpy-threshold pollutants, the GHG emission threshold is 75,000 tpy of $CO2_e$.

[h] If VOC emissions breach the exemption threshold, they are automatically considered to "significantly" affect onshore air quality.

End Notes

[1] Bureau of Ocean Energy Management, Regulation and Enforcement, "Offshore Production Facilities in Federal Waters," at http://www.boemre.gov/stats/PDFs/OCS PlatformActivity.pdf.

[2] For statistical information, see BSEE's "Borehole Online Query," at http://www.data.bsee.gov/homepg/data_center/ well/borehole/master.asp.

[3] After the 2010 Deepwater Horizon incident the MMS was replaced by the Bureau of Ocean Energy Management, Regulation and Enforcement (BOEMRE), which was subsequently divided into three separate entities: the Bureau of Ocean Energy Management (BOEM) and the Bureau of Safety and Environmental Enforcement (BSEE), and Office of Natural Resources Revenue (ONRR).

[4] See CRS Report R41132, *Outer Continental Shelf Moratoria on Oil and Gas Development*, by Curry L. Hagerty.

[5] For more information, see http://www.gomr.boemre.gov/homepg/lsesale/220/matl220.html.

[6] For more information, see CRS Report R41153, *Changes in the Arctic: Background and Issues for Congress*, coordinated by Ronald O'Rourke.

[7] CRS Report R41485, *Reorganization of the Minerals Management Service in the Aftermath of the Deepwater Horizon Oil Spill*, by Henry B. Hogue.

[8] See 76 *Federal Register* 64432 (October 18, 2011).

[9] S. Rept. 101-228 (October 27, 1990).

[10] See permit materials (e.g., EPA's Statement of Basis documents) from EPA Region 4 and Region 10, at http://www.epa.gov/region4/air/permits/OCSPermits/OCSpermits.html and http://yosemite.epa.gov/R10/ AIRPAGE.NSF/Permits/ocsap/. Accessed in September 2011.

[11] The CAA defines criteria pollutants as those that "endanger public health or welfare," and whose presence in ambient air "results from numerous or diverse mobile or stationary sources." Six pollutants are currently identified as such: ozone, particulates, carbon monoxide, sulfur dioxide, nitrogen oxides, and lead. See CRS Report R41563, *Clean Air Issues in the 112th Congress*, by James E. McCarthy.

[12] Lisa P. Jackson (EPA Administrator), Testimony before the House Committee on Oversight and Government Reform, May 24, 2011.

[13] May 22, 1953, 67 Stat. 29.

[14] August 7, 1953, 67 Stat. 464.

[15] 43 U.S.C. §1334 (from 1958 version).

[16] For more information, see CRS Report RL30798, *Environmental Laws: Summaries of Major Statutes Administered by the Environmental Protection Agency*, coordinated by David M. Bearden.

[17] Congress made major amendments to the CAA in 1977 and 1990. See CRS Report RL30853, *Clean Air Act: A Summary of the Act and Its Major Requirements*, by James E. McCarthy et al.

[18] CRS Memorandum, Jurisdiction of OCS Air Quality Control in 1978 OCSLAA, September 23, 1988.

[19] 43 *Federal Register* 16393 (April 18, 1978).

[20] 43 U.S.C. §§1331-1356. The OCSLA was originally enacted in 1953, but has been amended several times.

[21] DOI delegated most of this responsibility to the Minerals Management Service. After the 2010 Deepwater Horizon incident the MMS was replaced by the Bureau of Ocean Energy Management, Regulation and Enforcement (BOEMRE), which was subsequently divided into two separate entities: the Bureau of Ocean Energy Management (BOEM) and the Bureau of Safety and Environmental Enforcement (BSEE).

[22] *California v. Kleppe*, 604 F.2d 1187 (9th Cir., 1979). For a discussion of this case, see Leon Harmon, "Recent Developments: *California v. Kleppe*" in *Environmental Law*, Vol. 10, No. 3, Spring 1980 (Northwestern School of Law of Lewis and Clark College).

[23] National Research Council, *Assessment of the U.S. Outer Continental Shelf Environmental Studies Program: Social and Economic Studies,* 1992, p. 129.

[24] Clean Air Act Amendments of 1990 (P.L. 101-549); 42 U.S.C. §7401 et seq.

[25] Minerals Management Service, *Gulf of Mexico Air Quality Study*, Final Report, Volume I, 1995, at http://www.gomr.boemre.gov/PI/PDFImages/ESPIS/3/3424.pdf.

[26] 43 U.S.C. §§1301-1315.

[27] Most state waters extend 3 nautical miles (1 nautical mile = 6,076 feet, or 1.15 miles) from shore. Louisiana waters extend 3 imperial nautical miles (1 imperial nautical mile = 6,080 feet). Texas and Gulf Coast of Florida waters extend 3 marine leagues (equating to 9 nautical miles). See CRS Report RL33404, *Offshore Oil and Gas Development: Legal Framework*, by Adam Vann.

[28] 43 U.S.C. §1311.

[29] The United States declared its EEZ in Presidential Proclamation No. 5030, 48 *Federal Register* 10605 (March 14, 1983). This declaration is consistent with the United Nations Convention on the Law of the Sea (UNCLOS). UNCLOS provides a comprehensive international legal framework intended for building consensus on actions related to the world's ocean spaces, uses, and resources. See United Nations Convention on the Law of the Sea, opened for signature December 10, 1982, in force November 16, 1994, 1833 U.N.T.S. 396, reprinted in United Nations, the Law of the Sea: United Nations Convention on the Law of the Sea (UN Pub. Sales No. E.83.V.5). For additional information, see CRS Report RS21890, *The U.N. Law of the Sea Convention and the United States: Developments Since October 2003*, by Marjorie Ann Browne.

[30] 43 U.S.C. §1314.

[31] For more information, see CRS Report R41132, *Outer Continental Shelf Moratoria on Oil and Gas Development*, by Curry L. Hagerty; and CRS Report R40645, *U.S. Offshore Oil and Gas Resources: Prospects and Processes*, by Marc Humphries, Robert Pirog, and Gene Whitney.

[32] See CRS Report RL33404, *Offshore Oil and Gas Development: Legal Framework*, by Adam Vann.

[33] Most state waters extend 3 nautical miles (1 nautical mile = 6,076 feet, or 1.15 miles) from shore. Florida is the only state within EPA's jurisdiction with a different boundary: the Gulf Coast of Florida waters extend 3 marine leagues (equating to 9 nautical miles). See also CRS Report RL33404, *Offshore Oil and Gas Development: Legal Framework*, by Adam Vann. See the BOEMRE OCS website ("Definitions and Information") at http://boemre.gov/incidents/ def_info.htm.

[34] 42 U.S.C. §7627.

[35] 57 *Federal Register* 40792 (September 4, 1992).

[36] 56 *Federal Register* 63774 (December 5, 1991).

[37] CAA §§160-169.

[38] Such areas are termed "attainment areas." For more information, see CRS Report RL30853, *Clean Air Act: A Summary of the Act and Its Major Requirements*, by James E. McCarthy et al.

[39] 40 C.F.R. §52.21(b). In the CAA, the applicable term is "major emitting facility" (CAA §169).

[40] The official term is "regulated NSR [New Source Review] pollutant," which includes (among others) any pollutant for which a national ambient air quality standard (NAAQS) has been promulgated—sulfur dioxide (SO2), particulate matter (PM2.5 and PM10), nitrogen dioxide (NO2), carbon monoxide (CO), ozone, and lead—and any pollutant identified (in this citation) as a constituent or precursor for a regulated NSR pollutant (40 C.F.R. §52.21(b)(50)).

[41] CAA §169(1) lists 28 categories of sources with the 100 tpy threshold (also in 40 C.F.R. §52.21(b)). These sources include specific industries, such as cement plants and petroleum refineries, but not oil/gas exploration, development, or production activities.

[42] In the context of the PSD program, GHGs are the aggregate group of the following six gases: carbon dioxide (CO2) nitrous oxide (N2O), methane (CH4), hydrofluorocarbons, perfluorocarbons, and sulfur hexafluoride (SF6) (40 C.F.R. §52.21(b)(49)).

[43] For more information, see CRS Report R41212, *EPA Regulation of Greenhouse Gases: Congressional Responses and Options*, by James E. McCarthy.

[44] GHG emissions are often presented in carbon dioxide equivalents (CO2e), because GHGs vary by global warming potential (GWP). GWP is an index of how much a GHG may contribute to global warming over a period of time, typically 100 years. GWPs are used to compare

gases to CO2, which has a GWP of 1. For example, methane's GWP is 25, and thus a ton of methane is 25 times more potent a GHG than a ton of CO2.

[45] For more information, see U.S. EPA, "Prevention of Significant Deterioration and Title V Greenhouse Gas Tailoring Rule," 75 *Federal Register* 31514. A six-page EPA Fact Sheet summarizing the rule is available at http://www.epa.gov/ nsr/documents/20100413fs.pdf.

[46] 40 C.F.R. §52.21(j).

[47] The regulations list the "significant amounts" for each pollutant in 40 C.F.R. §52.21(b)(23).

[48] 40 C.F.R. §52.21(b)(12).

[49] For more information, see EPA Region 4, *Prevention of Significant Deterioration Permit Application Requirements,* (updated February 2010), which references EPA, *New Source Review Workshop Manual-Prevention of Significant Deterioration and Nonattainment Area Permitting*, Draft Guidance, 1990.

[50] For more information on NAAQS, see CRS Report RL30853, *Clean Air Act: A Summary of the Act and Its Major Requirements*, by James E. McCarthy et al.

[51] 40 C.F.R. §52.21(k).

[52] 40 C.F.R. §52.21(m).

[53] EPA has the discretion to shorten this time frame to a period not less than four months (40 C.F.R. §52.21(m)).

[54] 40 C.F.R. §52.21(o).

[55] 40 C.F.R. §52.21(e).

[56] 40 C.F.R. §52.21(p).

[57] 40 C.F.R. §52.21(m)(2).

[58] 40 C.F.R. §52.21(i)(5).

[59] 40 C.F.R. §55.8.

[60] 40 C.F.R. §52.21(i)(3).

[61] 45 *Federal Register* 52676 (August 7, 1980).

[62] As technology evolves, EPA adds new standards (per CAA §111). As of September 2011, 40 C.F.R. Part 60 contains over 90 standards.

[63] See e.g., EPA Region 4, Draft Permit to Shell Offshore Inc., at http://www.epa.gov/region4/ air/permits/OCSPermits/ OCSpermits.html

[64] For further discussion of CAA Section 112, see CRS Report RL30853, *Clean Air Act: A Summary of the Act and Its Major Requirements*, by James E. McCarthy et al.

[65] 40 C.F.R. §55.13(e).

[66] Although several permit documents identify Part 61, Subpart J (fugitive benzene emissions from specific equipment) as potentially applicable, none of the operations were ultimately subject to Part 61. See "Statement of Basis" documents from EPA Region 4 and Region 10, at http://www.epa.gov/region4/air/permits/OCSPermits/ OCSpermits.html and http://yosemite.epa.gov/R10/AIRPAGE.NSF/Permits/ocsap/. Accessed in September 2011.

[67] CAA Title V, which references CAA Section 302 and the definition of "major stationary source." This is codified in 40 C.F.R. part 71. The threshold for hazardous air pollutants is even lower: 10 tpy of an individual hazardous pollutant or more than 25 tpy of any combination.

[68] For more information, see CRS Report RL33632, *Clean Air Permitting: Status of Implementation and Issues*, by Claudia Copeland.

[69] C.F.R. §55.14.

[70] For more information, see CRS Report RL30853, *Clean Air Act: A Summary of the Act and Its Major Requirements*, by James E. McCarthy et al. See also,

[71] As mentioned above, a BACT determination may consider a proposed control measure's energy, environmental, and economic impacts.

[72] Although P.L. 112-74 (signed by President Obama December 23, 2011) transferred air emission authority in the OCS off Alaska's north coast to DOI, the act states the transfer "shall not invalidate or stay (1) any air quality permit pending or existing as of the date of the enactment of this Act; or (2) any proceeding related thereto" (Section 432).

[73] Of the recently EPA-issued OCS air permits, nitrogen oxide (NO_x) emissions were the primary reason an OCS source required a permit. Although the United States has no nitrogen dioxide (NO2) non-attainment areas (per EPA's Office of Air and Radiation website, at http://www.epa.gov/air), NO_x emissions are precursors for ozone and PM2.5.

[74] EPA's nonattainment map for ozone (*Figure 2*) is based on its 1997 ozone NAAQS. EPA lowered this standard from 0.08 parts per million (ppm) to 0.075 ppm in 2008, but nonattainment areas for the new standard have not yet been designated. See CRS Report R41062, *Ozone Air Quality Standards: EPA's Proposed Revisions*, by James E. McCarthy.

[75] 40 C.F.R. §55.4

[76] 40 C.F.R. §55.5.

[77] CAA §328(a)(4)(B), 42 U.S.C. 7627(a)(4)(B).

[78] 57 *Federal Register* 40795 (September 4, 1992).

[79] Some raised the issue again during Shell's air permit process for drilling in the Chukchi Sea. See EPA Environmental Appeals Board, Opinion Regarding Shell OCS Permit, Decided December 30, 2010.

[80] 42 U.S.C. §7627(a)(4)(C).

[81] 57 *Federal Register* 40792 (September 4, 1992). EPA's rationale for this interpretation relates to the second clause in the OCS definition: "is regulated or authorized under the [OCSLA]." According to EPA, the OCSLA (43 U.S.C. 1333) specifically excludes ships and vessels (unless permanently or temporarily attached to the seabed) from its regulatory authority.

[82] For example, see EPA permit (and associated documents) for Shell for oil exploration in the Chukchi Sea (issued September 19, 2011), at http://yosemite.epa.gov/R10/ airpage.nsf /Permits/chukchiap/.

[83] CAA §328(a)(3).

[84] 40 C.F.R. §55.11(a). This latter authority was not included in the initial final rule, but added by a subsequent rulemaking (62 *Federal Register* 46406, September 2, 1997).

[85] 40 C.F.R. §55.11(b).

[86] Provided in 40 C.F.R. §55.11(j).

[87] Santa Barbara County Air Pollution Control District (SBCAPCD); South Coast Air Quality Management District (SCAQMD); Ventura County Air Pollution Control District (VCAPCD); and San Luis Obispo County Air Pollution Control District (SLOAPCD).

[88] 73 *Federal Register* 66037 (November 8, 2008).

[89] 40 C.F.R. §124.10; 40 C.F.R. §71.11.

[90] 40 C.F.R. §124.11; 40 C.F.R. §71.11.

[91] 40 C.F.R. §124.12; 40 C.F.R. §71.11.

[92] 40 C.F.R. §124.13; 40 C.F.R. §71.11.

[93] 40 C.F.R. §124.14; 40 C.F.R. §71.11.

[94] 40 C.F.R. §124.19; 40 C.F.R. §71.11.

[95] Further information regarding the EAB is available at http://www.epa.gov/eab/.

[96] If a specific statute is silent, the Administrative Procedures Act provides for judicial review. For more information, see CRS Report R41546, *A Brief Overview of Rulemaking and Judicial Review*, by Vanessa K. Burrows and Todd Garvey.

[97] See CRS Report RL33632, *Clean Air Permitting: Status of Implementation and Issues*, by Claudia Copeland.

[98] "A petition to the Environmental Appeals Board under paragraph (a) of this section is, under 5 U.S.C. 704, a prerequisite to the seeking of judicial review of the final agency action" (40 C.F.R. 124.19(e)). For further information, see EPA, *The Environmental Appeals Board: Practice Manual*, September 2010.

[99] 43 U.S.C. §1334(a)(8).

[100] 45 *Federal Register* 15128 (March 7, 1980).

[101] 63 *Federal Register* 29487 (May 29, 1998).

[102] 76 *Federal Register* 64432 (October 18, 2011).

[103] 30 C.F.R. §250.103 (BSEE) and §550.103 (BOEM).

[104] 43 U.S.C. §1340.

[105] 43 U.S.C. §1351.

[106] 30 C.F.R. §250.201 (BSEE) and §550.201 (BOEM).

[107] BSEE reviews Deepwater Operations Plans (DWOPs), but these are supplemental to other activity-specific plans and do not have additional air emission provisions.

[108] 30 C.F.R. §550.218 (EP) or §550.249 (DPP or DOCC).

[109] MMS NTL No. 2009-N11, "Air Quality Jurisdiction on the OCS," effective December 4, 2009.

[110] 30 C.F.R. §550.303(d).

[111] The CO emission exemption formula is $E = 3400D^{2/3}$. Thus, a facility 50 miles from shore would be exempt if its projected CO emissions were less than 46,750 tpy.

[112] 30 C.F.R. §550.303(e).

[113] EPA levels are in 40 C.F.R. §51.165(b).

[114] 30 C.F.R. §550.303(g)(4).

[115] 30 C.F.R. §550.303(k).

[116] 30 C.F.R. §550.302. See also the 1980 *Federal Register* preamble that states: "The Department also believes that it is appropriate, particularly in the initial stages, for lessees to identify BACT." 45 Federal Register 15135 (March 7, 1980).

[117] 30 C.F.R. §550.303(g).

[118] 30 C.F.R. §550.303(g).

[119] 40 C.F.R. §52.21(c).

[120] For example, EPA replaced the TSP standard with the PM_{10} standard in 1987 (52 *Federal Register* 24634, July 1, 1987). See also http://www.epa.gov/pm/history.html.

[121] 30 C.F.R. §550.303(h).

[122] 30 C.F.R. §550.302.

[123] As stated in the introductory text of §328(a)(4), the definitions apply to §328(a), which discusses EPA's program, and §328(b), which covers the OCS areas adjacent to Texas, Louisiana, Mississippi, and Alabama.

[124] 30 C.F.R. §550.224 (EP) and §550.257 (DPP).

[125] 30 C.F.R. §550.218(a) for EPs; §550.249(a) for DPPs.

[126] DOI, "Gulf of Mexico Air Emission Calculations Instructions," accompanying Form BOEM-0138.

[127] 16 U.S.C. §1456(c).

[128] 30 C.F.R. §550.235

[129] See e.g., Sam Kalen, "The BP Macondo Well Exploration Plan: Wither the Coastal Zone Management Act?," *Environmental Law Reporter*, November 2010.

[130] National Oceanic and Atmospheric Administration, Ocean and Coastal Resource Management, Appeals to the Secretary of Commerce Under the Coastal Zone Management Act (CZMA) - January 15, 2009, http://www.coastalmanagement.noaa.gov/consistency/media/appealsdecisionlist011509.pdf.

[131] For more information, see CRS Report RL34339, *Coastal Zone Management: Background and Reauthorization Issues*, by Harold F. Upton; and CRS Report RL33404, *Offshore Oil and Gas Development: Legal Framework*, by Adam Vann.

[132] 15 C.F.R. Part 930.

[133] See 71 *Federal Register* 790 (January 5, 2006). See also NOAA, *CZMA Federal Consistency Overview*, February 2009.

[134] 30 C.F.R. §550.267.

[135] 43 U.S.C. §1349.

In: Federal Control of Air Emissions ISBN: 978-1-62081-808-4
Editors: H.M. Graham and T.E. Lee

Chapter 2

EPA'S BOILER MACT: CONTROLLING EMISSIONS OF HAZARDOUS AIR POLLUTANTS*

James E. McCarthy

SUMMARY

On December 2, 2011, EPA Administrator Lisa Jackson signed proposed revisions to EPA's recently promulgated Maximum Achievable Control Technology standards for boilers (the "Boiler MACT").

Publication of the proposed revisions in the Federal Register on December 23 began a 60-day public comment period, after which EPA will complete reconsideration of the rule—presumably by mid-2012. The Boiler MACT standards now being reconsidered were promulgated March 21, 2011, to meet the requirements of Section 112 of the Clean Air Act, but EPA stayed implementation of the requirements until it completes the reconsideration process.

There is widespread interest in the rule's requirements and their potential effects, because boilers are used as power sources throughout industry and for power or heat by large commercial establishments and institutions as well.

---

* This is an edited, reformatted and augmented version of a Congressional Research Service publication, CRS Report for Congress R41459, dated March 9, 2012.

EPA developed the regulations because it has found, based on emissions data, that boilers (including coal-, biomass-, and liquid-fired boilers) are major sources of hazardous air pollutants (HAPs).

The Clean Air Act defines a major source as any facility that emits 10 tons or more of a single listed HAP or 25 tons of any combination of HAPs annually. The HAPs themselves (187 substances) were listed by Congress in the 1990 Clean Air Act Amendments.

When finalized, the rule will replace a 2004 version of the rule that was vacated and remanded to EPA by the D.C. Circuit Court of Appeals. EPA has been under a court order to promulgate a replacement.

As proposed December 2, the MACT would affect about 14,000 boilers and process heaters, with capital costs of $5.4 billion, according to the agency; annualized costs, which spread the costs of capital over the expected life of the equipment and include operating and maintenance expenses, are estimated at slightly less than $1.5 billion per year. In response to comments on an earlier proposal and on the promulgated rule, the current EPA proposal would reduce the number of units expected to require controls, and make the emissions standards much less stringent, reducing the agency's estimate of annualized control costs from $2.9 billion to $1.49 billion.

Most of the costs would be borne by boilers that burn coal, biomass, or liquid fuels; only 12% of all the units covered by the rule will need to install equipment to meet it. Most of the boilers affected by the rule (83%) are fueled by natural gas or refinery gases. These boilers would not have to install pollution control equipment and most would experience cost savings under the rule, according to EPA. For the rule as a whole, EPA estimated that benefits—including the avoidance of 3,100 to 8,000 premature deaths annually—would outweigh costs by at least $25 billion per year.

Affected industries and many in Congress have raised objections to the rule as proposed and as promulgated, and bills have been introduced in both the House and Senate (H.R. 2250 and S. 1392) to alter the rule's requirements and delay its implementation. H.R. 2250 passed the House 275-142 on October 13.

Provisions similar to H.R. 2250 were offered as an amendment (S.Amdt. 1660) to the Senate version of the surface transportation bill (S. 1813) on March 8, 2012, but were not adopted.

In addition to the Boiler MACT, this report discusses three related rules that EPA promulgated at the same time, dealing with smaller "area source" boilers and with commercial and industrial boilers that burn solid waste (the "CISWI" and solid waste rules).

The latter two rules have also been controversial. Like the Boiler MACT, the CISWI rule was stayed by EPA, pending reconsideration.

INTRODUCTION

This report provides information concerning EPA's Maximum Achievable Control Technology standards for boilers (the Boiler MACT), an EPA rule designed to reduce emissions of hazardous air pollutants. On December 2, 2011, EPA proposed changes to the Boiler MACT standards that it had promulgated March 21, 2011.[1] The agency expects to finalize its reconsideration of the rule within the next few months, after which owners and operators of affected boilers would begin the process of obtaining or revising permits and installing equipment necessary to meet the new emission standards. The final standards are also likely to be challenged in the U.S. Court of Appeals for the D.C. Circuit, where numerous parties have already filed suit based on the March 2011 version of the rules.[2]

In addition to the Boiler MACT, this report discusses three related rules that were promulgated on March 21, 2011, and for which changes have also been subsequently proposed. The related rules set standards for small boilers ("area sources") and boilers that use solid waste as fuel ("commercial and industrial solid waste incinerators"), and they identify what materials EPA considers to be solid waste. Because boilers are used as power sources throughout industry, and for power or heat by large commercial establishments and institutions, there has been widespread interest in the requirements of all four of these rules and their potential effects.

Under the March 2011 rules, major source boilers subject to the MACT rule would have faced a May 2014 deadline for compliance. On May 16, 2011, however, EPA announced that it was staying the effective dates[3] of the Boiler MACT and one of the related rules, in order to take additional public comment and reconsider what it had promulgated, leaving in doubt both the final form of the rules and when the standards might go into effect. The federal District Court for the District of Columbia vacated the agency's stay on January 9, 2012,[4] but the court's action appears to have had little effect, as EPA subsequently issued a "No Action Assurance Letter." The letter states that the agency will exercise its enforcement discretion to not enforce certain notification deadlines in the March 2011 rule.[5] The agency also proposed, in its reconsideration of the rules, to extend the effective date (and hence the required compliance date) of the Boiler MACT until three years after completion of the rules' reconsideration.[6] Assuming the agency finalizes this proposal, it will have added at least 15 months to the date when emissions sources will have to comply.

A March 5, 2012 letter from EPA Administrator Lisa Jackson to Senator Ron Wyden elaborated further:

> In the proposal, EPA proposed to "reset" the three year compliance clock to give entities the full amount of time available under the Clean Air Act upon finalization of the rule, and, subject to the formal rulemaking process, expects to do so in the final rule. The Act also gives state and local permitting authorities the ability to provide up to a one-year extension of that deadline, on a case-by-case basis, as necessary, for the installation of controls.7

Thus, the agency envisions up to a four-year compliance deadline if sources can demonstrate that extra time is needed for the installation of controls.

Even before EPA's May 16 decision to delay the rules' effective date, the agency had announced its intention to reconsider the promulgated rules. The Notice of Reconsideration, which appeared in the March 21 *Federal Register*,[8] listed 14 provisions (including the subcategories the agency used to set standards and the emission standards themselves) for which the agency thought additional opportunity for public review and comment should be obtained. It stated that the agency might seek public comment on other aspects of the rules, as well.

The May 16 notice[9] established a 60-day public comment period, which ran through July 15, and it delayed the effective date of the rules until EPA completes its reconsideration of the rules or until proceedings for judicial review of the rules are complete, whichever is earlier. EPA subsequently announced that it expected to propose any changes to the rules by October 31, 2011, and finalize its decisions by April 30, 2012. The October deadline slipped to December 2. It would seem likely that the April 30 deadline for the final version of the reconsidered rules will also slip, as the agency must review and respond to public comments and undergo interagency review of the reconsidered regulations before they can be finalized.

Given the rules' potential impacts and the uncertainty of their final form, there continues to be concern among stakeholders as to the final requirements. EPA estimates that, as proposed on December 2, the Boiler MACT would affect about 14,000 boilers and process heaters.[10] In order to reduce emissions of a wide array of hazardous air pollutants, about 12% of the affected units would be required to install pollution control equipment. The 12% include coal-fired, biomass-fired, and liquid-fired boilers. The agency estimates the capital costs associated with the rule at $5.4 billion to meet the compliance deadline in 2015; annualized costs, which spread the costs of capital over the expected life of the equipment and include operating and maintenance costs as well, were estimated at $1.49 billion per year. These cost estimates are about half the estimated cost of EPA's originally proposed version of the rule, for reasons that we will explore below.

Most boilers—83% of those affected by the rule—are fueled by natural gas or similar gases such as refinery gas, according to EPA. These gas-powered boilers would incur capital costs averaging a little less than $6,500 per unit, according to the agency. Through fuel savings, the agency expects a reduction in operating costs to more than compensate for the capital expenditures of most gas-powered units.

WHY IS EPA CONSIDERING REGULATING THESE SOURCES?

EPA has developed regulations addressing boiler emissions because it has found, based on emissions data, that the coal-fired, biomass-fired, and liquid-fired units are major sources of hazardous air pollutants (HAPs). Section 112 of the Clean Air Act, which requires controls on major sources of HAPs, defines a major source as any facility that emits 10 tons or more of a single listed HAP or 25 tons of any combination of HAPs annually. The HAPs themselves (187 substances) were listed by Congress in the 1990 Clean Air Act Amendments.

Boilers emit at least 20 of the listed HAPs, including mercury, arsenic, chromium, cadmium, selenium, nickel, lead, manganese, phosphorous, antimony, beryllium, polycyclic organic matter, benzene, formaldehyde, acetaldehyde, dioxins, furans, hydrogen chloride, hydrogen cyanide, and hydrogen fluoride. Six of these 20 are classified as known or probable human carcinogens. Others affect the lungs, skin, central nervous system (including adverse developmental effects), and/or kidneys.[11] By controlling boiler emissions, EPA expects to avoid 3,100 to 8,000 premature deaths annually, as well as many other health effects, including 4,900 nonfatal heart attacks annually.[12]

The Boiler MACT will replace a rule promulgated on September 13, 2004, and subsequently vacated and remanded to EPA by the D.C. Circuit Court of Appeals.[13] The court vacated that rule in 2007, saying EPA had wrongly excluded many industrial boilers from the definition of solid waste incinerators, which have more stringent emissions requirements under the Clean Air Act. EPA was under a court order to finalize replacement rules by February 21, 2011.

Reconsideration of the Rule

In early December 2010, the agency petitioned the District Court for the District of Columbia for up to 15 months of additional time to complete the rulemaking. The agency argued that in light of the extensive comments it received on the proposed rules, "EPA believes that the overall public interest is best served by allowing EPA to re-propose the rules so that [it] will be able to issue emission standards that are based upon a thorough consideration of all available data and reduce potential litigation risks."[14] The court had issued a summary judgment against the agency in 2006 for failure to discharge fully its duty to promulgate standards for emissions of hazardous air pollutants.[15] On March 31, 2006, the court imposed a schedule under which EPA was to have discharged all of the statutory duties at issue by June 15, 2009. That deadline was subsequently extended by more than a year and a half.

On January 20, 2011, the court denied EPA's request for a further 15-month extension, concluding that EPA had engaged in discretionary delay in the face of a congressional directive (i.e., the 1990 Clean Air Act Amendments, under which the rules were to have been promulgated by November 2000); the court gave the agency one month to issue final rules.[16] Having been denied the extension it sought, the agency issued a statement saying, "The standards will be significantly different than what EPA proposed.... The agency believes these changes still deserve further public review and comment and expects to solicit further comment through a reconsideration of the rules."[17]

True to its word, the agency issued a Notice of Reconsideration at the same time that it promulgated the final rule. As noted earlier, the notice listed 14 provisions for which the agency thought additional opportunity for public review and comment should be obtained, and it stated that the agency might seek public comment on other aspects of the rules. The 14 provisions included such basic elements as the subcategories used to set standards in the boiler rule, the emission standards themselves, and the monitoring requirements. This left numerous questions concerning not only the substance of the rule, but the schedule for implementation. Following promulgation, existing facilities would normally have three years to comply with the standards, but if the agency is reconsidering key aspects of the rules, one was left to wonder how regulated entities should determine what standards they will ultimately be required to comply with, and on what schedule. By staying the effective date of the standards on May 16, 2011, the agency addressed this uncertainty, effectively giving itself and regulated entities an extension of time similar to what the court had denied them.

The Promulgated Standards

Standards for Existing Coal, Biomass, and Liquid Units

In order to understand the standards that EPA proposed December 2 and the controversy surrounding them, it helps to begin with the agency's June 4, 2010, proposal (which we refer to, generally, in this report as the standards EPA "originally" proposed). In that proposal, EPA divided boilers into 11 subcategories, with separate emission limits for new and existing units in 9 of the 11.[18] The nine subcategories included three types of coal-fired boilers and four types of biomass-fired boilers.[19]

The proposed emission limits covered five substances (or groups of substances): mercury; dioxins/furans; particulate matter (as a surrogate for non-mercury metals); hydrogen chloride (as a surrogate for all acid gases); and carbon monoxide (as a surrogate for non-dioxin organic air toxics, including formaldehyde).[20] As explained below, EPA is now proposing to replace the emission standards for dioxins/furans with a work practice standard; if the agency finalizes this proposal, the emission standards would only address four groups of pollutants.

The Clean Air Act requires that MACT emission standards be based on the emission control achieved by the best controlled similar sources. Thus, the emission limits originally proposed for the five groups of pollutants were based on monitoring data obtained from facilities in each of the nine subcategories of existing boilers.[21]

- For *new* sources, the statute requires (in §112(d)(3)) that standards be based on the emission control achieved by the best controlled similar source.
- For *existing* sources, on the other hand, the same subsection of the statute
- requires standards no less stringent than the average emission limitation achieved by the best performing 12% of existing sources. The performance of the best 12% is generally referred to as the "MACT floor," since it sets the minimum requirements for MACT

> standards. The MACT floor is based solely on the performance of existing facilities in the category or subcategory of sources, with no consideration of the cost or economic impacts thereof. The Administrator is only allowed to take costs, health, energy, and environmental factors into consideration to the extent that she considers setting standards that go beyond the floor.

Given the statutorily required methodology for identifying the MACT floor, the number of subcategories the agency identifies is an important factor in determining how stringent the standards will be: the more subcategories EPA identifies, the more variation there will be in the MACT floor for each, and thus the more flexibility the agency will have in setting different, potentially less stringent emission standards for different boiler types. If, because of subcategorization, the Administrator decided that a subcategory's MACT floor did not provide sufficient protection for human health or the environment, she would still have the authority to set "beyond the floor" standards for a subcategory: in doing so, however, she could consider the cost of the standards and other factors. Thus, one issue raised by commenters on the proposed rule was whether EPA's subcategorization of the boiler universe appropriately considered the differences in size, fuels, boiler design, location, etc., or whether the subcategories should be modified from those originally proposed.

In the final rule, EPA responded to the comments it received by modifying its subcategories. Instead of 11 types of boiler, the final rule identified 15, including an additional biomass subcategory, a limited-use subcategory,[22] and a subcategory for "non-continental" (i.e., Hawaiian, Puerto Rican, and U.S. island territories) liquid-fueled boilers.[23] Based on new data provided by industry and on some corrections to the data it had used earlier, the emission standards for almost every subcategory were modified, mostly making the standards less stringent.

In the re-proposal signed by the Administrator December 2, 2011, three additional subcategories were added, bringing the total to 18. The new proposal adds two biomass-fueled subcategories (bringing the number of biomass subcategories to seven) and divides liquid-fueled boilers into those that burn heavy-liquid fuel and those burning light liquid fuel. The re-proposal also sets new emission limits for the pollutants emitted from each of the subcategories.

The net effect of all these changes is a substantial easing of the proposed standards' stringency. As shown in *Table 1*, existing coal-fired boilers will be allowed 40% to 340% more particulate matter (PM) depending on boiler type, 10% more hydrogen chloride (HCl), and 3% more mercury than would have been allowed by the original June 2010 proposal. The increase in allowable emissions is even greater for most of the pollutants emitted by existing biomass units: they will be allowed to emit more than triple the HCl and mercury that would have been allowed under the originally proposed standards, and as much as 22 times the PM originally proposed. The standards for liquid-fueled units also changed, allowing six times as much mercury and at least seven times the carbon monoxide.

Of the changes provided by the most recent proposal, the dioxin/furan standards are perhaps the most significant. EPA determined that the dioxin limits in the March 2011 standards had been set below levels that could be accurately measured. As a result, the agency proposes to replace the numeric emission limits that it had set in March 2011 with work practice standards: the latter will require an annual tune up of the boiler to ensure good combustion, which EPA believes will ensure minimal dioxin emissions without requiring emissions monitoring.

A second issue raised by critics of the agency's proposal had to do with the way that EPA identified the best performers within the subcategories. As it has done previously for other categories of sources, EPA has averaged the emissions performance of the top 12% of existing units separately for each of the four pollutants subject to emission limits. Critics who believed the proposed standards were too stringent argued that by considering the pollutants separately, the agency was, in effect, cherry-picking the best performers and setting a combined standard for the pollutants that no existing facility may actually meet.

This question—whether one identifies the best-performing sources pollutant-by-pollutant or for all the pollutants as a group—was addressed in regard to another standard, the Hospital/Medical/Infectious Waste Incinerator rule, which EPA promulgated in October 2009.[24] In promulgating that rule, the agency stated:

> There is no reason not to consider emissions data and controls in use at sources that may be the best performers from some pollutants but not for other pollutants. The MACT floor controls applicable for one pollutant do not preclude the use of MACT floor controls for another pollutant. Therefore, it is appropriate to consider controls at sources employing MACT controls for some pollutants, but not all.25

Table 1. Change in Emission Standards for Existing Major Source Boilers, December 2011 Re-Proposed Boiler MACT vs. June 2010 Proposed Rule (% change in proposed emission standards)

Subcategory	Estimated Number of Units	Particulate matter (PM)	Hydrogen chloride (HCl)	Mercury	Carbon monoxide
Coal					
Coal Stoker	616	+40%	+10%	+3%	+340%
Coal Fluidized Bed		+340%			+87%
Pulverized Coal		+120%			-54%
Biomass					
Biomass – Wet Stoker / Sloped Grate / Other	508	+45%	+267%	+244%	+41%
Biomass – Kiln-Dried Stoker / Sloped Grate / Other		+1,500%			-55%
Biomass Fluidized Bed		+450%			+48%
Biomass Suspension Burner		+155%			-94%
Biomass Dutch Ovens / Pile Burners		+80%			-20%
Biomass Fuel Cells		+65%			+455%
Biomass Hybrid Suspension Grate		+2,100%			+286%
Liquid					
Heavy Liquid	947	+1,450%	+33%	+550%	+900%
Light Liquid		-15%			+600%
Non-continental Liquid		+100%			+1,700%
Gas					
Other Process Gases (Gas 2)	129	-87%	+56,566%	+3,850%	+300%

Source: U.S. EPA, with calculations performed by CRS.

Notes: In addition to these changes to the numeric emission standards, the agency's December 2011 proposal would replace the emission standards for dioxins and furans with a work practice standard.

EPA acknowledged in the preamble to that rule that "there appears ... to be a substantial ambiguity in the statutory language about whether the MACT floor is to be based on the performance of an entire source or on the performance achieved in controlling particular hazardous air pollutants."[26] But the agency noted that commenters in the past have not objected to the use of the pollutant-by-pollutant approach. They also noted that the D.C. Circuit

Court of Appeals has reviewed MACT floor determinations made on a pollutant-by-pollutant basis without finding error in the approach.[27] Thus, the agency believes the best reading of the act is that the standards are to be set on a pollutant-by-pollutant basis—the only exception being if there is reason to believe that control of one pollutant will lead to increased emissions of another.

In the preamble to the March 2011 Boiler MACT standards, EPA provided a similar discussion, concluding that, although the language of Section 112(d)(3) is ambiguous, "EPA's HAP-by-HAP approach fulfills the evident statutory purpose and is supported by the most pertinent legislative history."[28]

Standards for Existing Gas-Fired Units

In the final and re-proposed rules, two subcategories are not subject to emission limits: natural gas/refinery gas/clean gas (a subcategory that EPA calls Gas 1) and limited-use units (defined as units that operate less than 10% of the time). For these two, the agency set only a work practice standard, requiring that boilers be tuned up annually and that the owners submit reports to EPA setting forth specific information from the tune-up procedure. The Administrator has authority to substitute a work practice standard for emission standards when, in her judgment, it is not feasible to prescribe or enforce an emission standard. As noted earlier, 83% of existing boilers fall into the natural gas/refinery gas/other clean gas subcategory, and thus are only subject to the tune-up requirements. Limited use boilers that burn liquids would also qualify for work practice standards, requiring a tune-up every two years.

All boilers would also be required to perform a one-time energy assessment to identify cost-effective energy conservation measures.

Standards for New Boilers

EPA also promulgated MACT standards for new (as opposed to existing) major source boilers. These standards are, in all cases, more stringent than the standards for existing units—in many cases, substantially so.[29] The agency assumes that no new coal-fired major source boilers (and very few major source boilers of any kind) will be built in the next three years. The agency has stated that the projected type and number of new boilers comes from the

Energy Information Administration at the Department of Energy and is not based on the Boiler MACT.[30]

Of the estimated 1,844 new units, the agency expects 1,762 to be powered by natural gas, with annualized costs of compliance averaging $2,900 apiece. The other 82 new boilers are projected to be fueled by biomass, with annual compliance costs of $1.2 million each.[31]

EPA'S ESTIMATES OF THE BOILER MACT'S COSTS AND BENEFITS

Among the boilers affected by the Boiler MACT rule, there are an estimated 947 units that burn liquids and 1,124 units that burn solids (616 of them coal-fired and the rest biomass-fired). The rule also applies to other types of boilers, but these 2,071 units, which account for 14.4% of the affected units, account for 100% of the compliance cost for existing units.

In general, the promulgated emission limits apply to boilers that have a designed heat input capacity of 10 million Btu per hour or greater. How big is this? A coal-fired boiler subject to the MACT would be one that is capable of burning roughly 1,000 pounds (a half-ton) of coal per hour.[32] Wood has less energy per pound than coal: a biomass-fired boiler burning wood might require as much as 1,500 pounds of wood per hour to produce 10 million Btus. A boiler burning fuel oil would need to burn about 70 gallons per hour. Many of the boilers to be regulated are substantially larger than this. An analysis released by the Council of Industrial Boiler Owners (CIBO), for example, used a 250 million Btu/hour boiler as the base for its cost estimates.[33] For a boiler burning fuel oil, this would mean burning 1,750 gallons per hour.

In order to comply with the rule's emission limits, these facilities may need to install fabric filters (also known as baghouses) to achieve PM and mercury control; wet scrubbers to meet the hydrochloric acid limits; replacement burners, tune-ups, and combustion controls for carbon monoxide and organic HAPs; and carbon injection for mercury control. These are the available technologies for maximum control of the relevant emissions.

Some observers maintain that, because EPA weakened the standards as compared to what it originally proposed, the vast majority of facilities won't have to install these technologies. The National Association of Clean Air Agencies (NACAA), the association that represents state air pollution control officials, surveyed its members in 2008 to determine what should be defined as MACT. Using the data it obtained from state officials, NACAA concludes that EPA's final mercury emission standard for coal-fired boilers is almost 16 times higher than the average of the best performing 12%; the carbon monoxide standard is 213 times what the MACT floor should be, according to NACAA. The problem according to NACAA's Executive Director, is that "Compliance test results provided by state and local permitting officials were not used [in setting the MACT standards]; instead EPA relied on industry data."[34]

EPA explains that some of the data that NACAA provided could not be used, because the test reports were incomplete. The agency also notes that its process for setting a standard is more complicated than simply averaging the best test results. Specifically, the agency subjects the emissions data to what is called a "variability analysis." This type of analysis attempts to recognize that operating conditions and resulting emissions vary over time, yet facilities need to be in compliance with emissions limits at all times. Emissions can change for several reasons: there is variation in the amount of contaminants in fuel, for example; the boiler will sometimes be operating at less than full load; and statistical tests applied to the data are used to set the actual standard. The agency first identifies the best 12% by ranking the units based on their best test results. In the next step they add all available test results for those units. Finally, using a statistical test, they calculate a standard that these units can meet 99% of the time, despite variability in operating conditions. This results in standards that are less stringent than the straight average of the best 12% test results.[35]

EPA's Projected Costs

As shown in *Table 2*, EPA estimates the capital costs of the re-proposed rule to be $5.4 billion, with annualized costs of nearly $1.5 billion.[36] These costs fall almost entirely on units burning solids (coal or biomass) or liquids. Most boilers, which are fueled by natural gas, will experience a reduction in operating costs that more than compensates for any capital costs, according to EPA.

Despite the clear advantage that the promulgated rule would give to natural-gas-fired boilers, EPA did not consider fuel-switching as a potential compliance strategy, for a variety of reasons. In the preamble to the originally proposed rule, the agency stated: "This decision was based on the overall effect of fuel switching on HAP emissions, technical and design considerations discussed previously in this preamble, and concerns about fuel availability."[37] Although switching from solid to gaseous fuels "would decrease PM and some metals emissions, emissions of some organic HAP (e.g., formaldehyde) would increase,"[38] according to the agency's analysis. Further, the agency maintained, natural gas may be unavailable:

> Natural gas pipelines are not available in all regions of the U.S., and natural gas is simply not available as a fuel for many industrial, commercial, and institutional boilers and process heaters. Moreover, even where pipelines provide access to natural gas, supplies of natural gas may not be adequate.[39]

Table 2. Estimated Costs to Existing Boilers for Compliance with EPA's Re-Proposed Boiler MACT

Subcategory	Estimated Number of Affected Units	Capital Costs ($ million)	Annualized Cost ($ million)	Annualized Cost per Unit ($ thousand)
Coal units	616	$2,713	$953	$1,550
Biomass units	508	$639	$169	$333
Heavy-liquid units	322	$769	$264	$820
Light-liquid units	581	$930	$277	$477
Non-continental liquid units	44	$181	$42	$955
Gas 1 (natural gas / refinery gas / other clean gas) units	11,911	$77	($295)	($25)
Gas 2 (other) units	129	$132	$55	$426
Total	14,111	$5,441	$1,465	—

Source: U.S. EPA, Proposed Amendments to Boiler MACT Final Rule, Table 5, pre-publication copy, pp. 135-136.

Notes: (1) Parentheses indicate lower costs, resulting from fuel savings. (2) Per unit cost was calculated by CRS. Some of the difference in unit costs could be accounted for by differences in boiler size.

Nevertheless, if the cost of compliance is sufficiently great, the incentive to explore fuel-switching would seem substantial, particularly for facilities not burning a byproduct of the plant's operation. Recent accounts of the substantial increases in gas reserves as shale gas resources are developed could ease some of the natural gas availability concerns, and might bear further analysis.[40]

EPA used a multi-market partial equilibrium model developed for its Office of Air Quality Planning and Standards to determine how stakeholders in 100 U.S. industries might respond to the promulgated rule. The model found no U.S. industry in which production would decline by more than 0.05%.[41]

EPA's Projected Benefits

EPA estimates that implementation of the Boiler MACT, as promulgated, would reduce nationwide emissions from major source boilers and process heaters by

- 0.5 to 1.8 tons per year (tpy) of mercury,
- 2,200 tpy of non-mercury metals,
- 37,000 tpy of hydrogen chloride,
- 41,200 tpy of particulate matter (PM),
- 4,700 tpy of volatile organic compounds, and
- 558,430 tpy of sulfur dioxide.[42]

For most of these pollutants, the expected reductions are similar to those of the originally proposed rule. This is not the case for mercury, however. The June 2010 proposed version of the rule was estimated to reduce mercury emissions by 7.5 tons, at least four times as much as the current proposal. Boilers are currently thought to be the fourth-largest stationary source of mercury, yet other categories of sources have been required to reduce mercury emissions to a greater extent than will be required by the promulgated or re-proposed Boiler MACT rule.[43] EPA's explanation for the continued mercury emissions is that much of the remaining mercury comes from small oil-fired boilers, which do not currently have controls, and which individually emit relatively small amounts of mercury. Thus, when the agency defined MACT for these units, it did not result in substantial mercury reductions.[44]

According to EPA, beginning in 2014, emission reductions resulting from the rule would lead to important health benefits, including the annual avoidance of:

- 3,100 to 8,000 premature deaths,
- 2,000 cases of chronic bronchitis,

- 4,900 nonfatal heart attacks,
- 5,350 hospital and emergency room visits,
- 4,600 cases of acute bronchitis,
- 390,000 days when people miss work,
- 51,000 cases of aggravated asthma, and
- 96,000 cases of respiratory symptoms.[45]

EPA estimates the annual value of these benefits to range from $27 billion to $67 billion in 2015—outweighing the annualized costs by at least $25 billion. In its Regulatory Impact Analysis, the agency states that it was only able to provide a partial estimate of the value of the rule's benefits:

> We were unable to monetize the direct benefits associated with reducing HAPs in this analysis. In Section 7.5.5 of this RIA, we provide a full qualitative discussion of the direct health benefits associated with the reductions in emissions of HAPs anticipated by these rules, including a full discussion of the complexity associated with monetizing HAP benefits. We also provide maps of reduced mercury deposition in that section. Therefore, all monetized benefits provided in this analysis only reflect improvements in ambient PM2.5 and ozone concentrations. Thus, the monetized benefits estimate is an underestimate of the total benefits. The extent of this underestimate, whether small or large, is unknown.46

OTHER COST ESTIMATES

Not surprisingly for a rule of this size, EPA's cost estimate is not the only one available. Industry-funded studies of the originally proposed rule, including one from the Council of Industrial Boiler Owners (CIBO), placed the costs of the rule substantially higher than EPA's estimate, while an analysis by the National Association of Clean Air Agencies concluded that CIBO's study exaggerated the potential costs.

The CIBO Study

CIBO's study concluded that capital costs of the proposed Boiler MACT would be $20.7 billion, more than double EPA's cost estimate for the rule as originally proposed, and four times EPA's estimated cost of the rule as promulgated and re-proposed.[47] CIBO estimated the cost of carbon monoxide controls at $2.7 billion, 200 times EPA's estimate; and the report estimated the cost of carbon injection at $1.7 billion, 180 times the EPA amount. CIBO's estimate for hydrogen chloride controls was only three times as expensive as EPA's estimate; but, because EPA's estimate was already more than $3 billion, the difference would add $6 billion to the total cost of the rule.

CIBO identified six factors that accounted for most of the difference. According to the CIBO study:

- EPA used outdated control cost estimates;
- to achieve the proposed carbon monoxide (CO) limits, it would have been necessary to implement combustion controls, fuel feed system improvements, or install a CO catalyst, at far greater cost than EPA's conclusion that a tune-up or burner replacement would be adequate to achieve the CO limits;
- activated carbon injection, in addition to fabric filters, would be required to achieve the proposed rule's mercury limits at a far higher number of boilers, and EPA underestimated the cost of this technology by a factor of 15;
- PM emission controls would require fabric filters, which are more expensive than EPA's assumption that electrostatic precipitators (ESP) would be adequate to meet the standard;
- more expensive scrubbers than EPA identified would be required for hydrogen chloride control; and
- more facilities than EPA estimated would need to control dioxin/furan emissions.

Using an estimate of 16,000 jobs "at risk" of being eliminated for each billion dollars of cost, CIBO concluded that 337,702 jobs would be at risk from implementation of the proposed rule.[48] Nearly 70,000 of these would be in the directly affected industries; the rest would be in supplier industries or spread across local economies through reductions in spending by workers directly and indirectly affected.

Like the cost estimates, CIBO's jobs "at risk" were much higher than EPA's estimates. EPA found that the employment impact of the proposed rule would range from a loss of 6,000 jobs to a gain of 12,000.[49] For the re-proposed rule, EPA estimates employment impacts as ranging from a loss of 3,000 to a gain of 6,300, with a central estimate showing an increase of 1,600 jobs.[50]

Several factors help explain why CIBO's estimates were so much higher than EPA's. The most important was the choice of model used to estimate the economic impacts. CIBO used an input-output (I-O) model. I-O models are deterministic, in the sense that they start with an assumption regarding a change in one industry's output and track the effect of the change on other industries' output and employment. An assumed loss in one industry translates directly to bigger losses in the economy as a whole.

In the CIBO study, the assumption was that the expenditure of $20.7 billion on pollution controls would be equivalent to a reduction of output by that amount.[51] In fact, spending on pollution control does not cause an equal reduction in output. Rather, changes in output caused by pollution control expenditures would include increases in some industries along with declines in others.

- Output declines occur in cases where industries increase prices to cover their higher costs, and consumers respond by demanding less of the affected product. In these cases, the higher costs would generally reflect "annualized costs," not the full capital cost used by CIBO.
- Meanwhile, output in other industries will usually increase. As EPA noted, in the Administrator's September 28, 2010, letter to Senator Landrieu, expenditures on pollution control are not simply a loss to the economy: they stimulate demand and provide jobs in the pollution control sector.[52]

Thus, the assumption that output would decline by $20.7 billion at the base of CIBO's analysis is flawed. As a result, little credence can be placed in CIBO's estimate of job losses.

The NACAA Critique

An analysis by the National Association of Clean Air Agencies[53] cited other flaws in the CIBO study. In addition to echoing the critiques above—that CIBO included no estimates of economic or health benefits and treated one-time costs as recurring expenses—NACAA raised two other major points. First, NACAA maintained, CIBO exaggerated the cost of the proposed rules by overestimating the number of sources that must be controlled: CIBO's estimate of the number of sources that must be controlled is "grossly in error,"[54] according to NACAA, because CIBO assumed that any source for which there were no emissions data would have to install controls. NACAA collected stack test data for boilers in 2008 in order to develop a model Boiler MACT permit for use by the states, and thus is the source of much existing emissions information. Using its existing data sets, NACAA looked at coal-fired boilers and found that 87% of the 39 units for which there were emissions data already met EPA's proposed standard for carbon monoxide. In developing its cost estimates for the coal-fired units, however, CIBO assumed that none of the remaining 146 (untested) units in the subcategory would meet the proposed standard, and thus they would all have to install additional controls. Second, NACAA concluded that the available data "reveal a continuum of emissions performance where there are substantial numbers of units whose emissions are within 10 to 40 percent of the proposed standard." For many of these units "it should be anticipated that minor changes, such as blending in small amounts of clean fuel, will suffice in lieu of major capital projects."[55]

Furthermore, according to NACAA, EPA data show a large number of sources that are capable of burning a variety of fuels. This creates "significant opportunities to reduce capital costs by shifting the mix of fuels combusted," according to NACAA, a lower-cost compliance strategy that both CIBO and EPA did not consider. NACAA provided numerous examples of how such strategies might be used by different types of boilers.[56]

The AF&PA Study

The American Forest and Paper Association (AF&PA) also sponsored a report on the proposed rule's potential impacts, and subsequently updated the report to reflect the promulgated rule.[57] The updated AF&PA study concluded that the Boiler MACT would cost pulp and paper mills $2.85 billion in capital costs, and would place 36 mills at risk of closure, threatening 20,541 jobs.[58] Although the updated study estimated the costs of the rule to be 17% ($800 million) less than their estimate for the proposed rule, the authors increased their estimate of the number of plants at risk of closure and the associated job losses by 20% and 22% respectively. The updated study does not provide a direct explanation for why a smaller increase in cost would lead to greater plant closures and job losses.[59]

Some General Thoughts on Cost and Economic Impact

In general, over the last 40 years, Clean Air Act rules have proven less expensive than both EPA and industry estimates have projected before they were promulgated. As the EPA Administrator noted in a September 2010 speech, after recounting examples of exaggerated projections of the consequences of proposed rules, "the Clean Air Act has not only reduced harmful pollution—it has also been particularly effective at proving lobbyists wrong."[60] The National Association of Clean Air Agencies reached the same conclusion: "Pre-regulation estimates by industry sources have historically overstated the cost of compliance with proposed regulations, often by substantial amounts." The NACAA report cites the Clean Air Act's acid rain program, catalytic converters on automobiles, the removal of lead additives from gasoline, the replacement of ozone-depleting substances in air conditioners, and the impacts of the 1997 National Ambient Air Quality Standards for ozone and particulate matter as examples of major regulatory programs whose costs were overestimated.[61]

More to the point, as noted earlier, EPA legally cannot take cost or economic impact into consideration in identifying the MACT floor, and all of the numeric standards in the rule are based on the agency's determination of the MACT floor.

But the agency can distinguish among classes, types, and sizes of sources within categories or subcategories. This has led the agency to propose generally less stringent emission standards as the agency has identified additional subcategories from within the boiler universe. In response to a September 24, 2010, letter sent by 41 Senators, the Administrator stated that it was the agency's intent to "focus on making the regulatory subcategories appropriately reflect industrial variation in the real world, and on aligning the standards in each subcategory with the performance that real-word conditions prove are already achievable."[62] The Administrator explained that this would be possible because the affected companies and institutions had provided additional information in response to the EPA proposal.

The agency should not expect an entirely free hand in setting additional subcategories (or perhaps, even in promulgating standards based on the many subcategories it originally proposed).

In comments on the originally proposed rule, a group of four environmental organizations that frequently have challenged EPA regulations, objected to EPA's proposed subcategorization, calling it "unlawful, arbitrary, and unsupported by the record."[63] They note that while the act provides that the Administrator *may* distinguish among classes, types, and sizes of sources within a category, such subcategorization is not *required*: "the plain text of the Act demonstrates that Congress intended EPA to creates [sic] categories and subcategories as a step towards establishing emissions standards, ... not as part of a scheme to provide incentives for existing sources to avoid standards. And yet, that is the effect of EPA's subcategorization scheme."[64]

NACAA's comments also argue that EPA set several of the standards at levels that were more lenient than the MACT floor. The NACAA study provided details on two of these, the carbon monoxide and mercury standards for coal-fired boilers.[65]

SHOULD EPA HAVE SET HEALTH-BASED STANDARDS UNDER SECTION 112(D)(4)?

According to EPA, "emissions data collected during development of the proposed rule show that hydrogen chloride [HCl] emissions represent the predominant HAP emitted by industrial, commercial, and institutional (ICI) boilers, accounting for 61 percent of the total HAP emissions."[66] Given the importance of HCl emissions, one of the key issues in considering EPA's proposal was whether the agency should have exercised its authority to set standards for HCl and other acid gases under Section 112(d)(4), which gives the Administrator flexibility to set standards less stringent than MACT for HAPs that have a health threshold (i.e., substances that are not harmful to people exposed to levels below some threshold).

In developing and promulgating other regulations, including the vacated 2004 MACT standard for boilers, EPA established that HCl has a health threshold, that it is not classified as a human carcinogen, and that there is limited health risk associated with HCl emissions from discrete units. Nevertheless, in the June 2010 proposal (and in the current re-proposal), the Administrator decided not to exercise her discretion to set less stringent standards for HCl emissions for several reasons, including

1) the agency lacked information on the peak short-term emissions of HCl from boilers and thus could not determine whether acute exposures will pose health concerns;

2) HCl emissions from boilers mix with other emissions that are respiratory irritants, and EPA has no studies explicitly addressing the toxicity of these mixtures;
3) in considering whether to exercise her discretion under Section 112(d)(4), the Administrator must determine that a health-based standard in lieu of a MACT will not result in adverse environmental effects. HCl gas forms an acidic solution in the atmosphere and could exacerbate the impacts of acid deposition from sulfur and nitrogen oxides;
4) the agency had limited information on facility-specific emissions that it would need to set a health-based standard;
5) the agency would have needed to decide whether it would be appropriate to set 112(d)(4) standards for each acid gas emitted by boilers, or a single standard as a surrogate for them all; and
6) as proposed (and as promulgated), the MACT standard would result in significant reductions in emissions of other pollutants, most notably sulfur dioxide,
7) particulate matter, other acid gases, mercury, and other metals. These reductions would provide substantial public health benefits that would be foregone if the agency set a less stringent standard.[67]

Whether the agency should have set standards for HCl under Section 112(d)(4) was one of the key points raised in comments, including those submitted by 41 Senators in a letter to the Administrator, on September 24, 2010, and by 105 Members of the House in a letter submitted August 2, 2010. As the Senate letter stated:

> To help reduce the burden of the rule in a manner that does not compromise public health and safety, ... we ask that you carefully consider the extensive record that supported the Agency's determination to include health-based emissions limitations for hydrogen chloride and manganese in the previous Boiler MACT rulemaking that was set aside by the reviewing court on wholly unrelated grounds.68

In the March 2011 rule and in the December 2 re-proposal, the agency did not change its mind on the use of Section 112(d)(4), but it did significantly change the hydrogen chloride standards, presumably based on new data supplied by affected entities. The changes, which are summarized in *Table 3,* allow more HCl emissions from all types of sources, especially from biomass- and gas-fired units.

Table 3. Proposed and Promulgated Emission Limits for HCl Emissions from Existing Boilers (lbs./million Btu)

Type of Boiler	Proposed (June 2010) Limit	Promulgated (March 2011) Limit	Re- Proposed (December 2011) Limit	% Difference (December 2011 vs. June 2010)
Coal	0.02	0.035	0.022	+10%
Biomass	0.006	0.035	0.022	+267%
Liquid	0.0009	0.00033	0.0012	+33%
Gas 2 (other process gases)	0.000003	0.0017	0.0017	+56,566%

Source: U.S. EPA, Proposed and Final Boiler MACT Federal Register notices, Table 1, and Proposed Reconsideration notice, Table 1.

SMALLER (AREA) SOURCES

Smaller boilers (those at facilities that emit less than 10 tons of an individual HAP and less than 25 tons of all HAPS combined) face regulations as well, but for the most part the Clean Air Act allows them to meet a less stringent standard, termed "Generally Available Control Technology" (GACT). A separate rule setting standards for these "area sources" was promulgated the same day as the MACT standards, and these standards were also re-proposed December 2.[69] Although EPA is reconsidering this rule, it did not extend the rule's effective date.

The area source rule distinguishes boilers that have a heat input capacity of 10 million Btu per hour or more from those that are smaller. The smaller units make up the overwhelming majority of the units covered by the area source rule; they would be subject to GACT. Under GACT, these units would not be required to meet emission limits. Rather, they would be required to meet a work practice standard by performing a boiler tune-up every two years. According to EPA, "By improving the combustion efficiency of the boiler, fuel usage can be reduced and losses from combustion imperfections can be minimized. Minimizing and optimizing fuel use will reduce emissions of mercury and all other air toxics."[70]

Some units under the area source rule would be subject to MACT for at least some pollutants. These are the coal-fired units that have a heat input capacity of 10 million Btu per hour or more, but are at *facilities* that don't meet the major source definition because, even counting their boiler emissions, they emit less than 10 tons of any individual HAP and less than 25 tons of any combination of them. According to the agency, these larger boiler units at area sources would need to meet standards based on MACT for some of the pollutants they emit: "The final standards for existing and new coal-fired boilers at area sources are based on MACT for mercury and CO, and on GACT for PM. The final standards for existing and new biomass boilers and existing and new oil-fired boilers at area sources are based on GACT."[71]

Table 4. Annualized Compliance Costs for Area Source Boiler Rule

Source	Subcategory	Estimated Number of Affected Units	Annualized Cost ($ million)	Annualized Cost per Unit ($ thousand)
Existing Units	Coal	3,710	$37	$10
	Biomass	10,958	$24	$2
	Oil	168,003	$374	$2
New Units	Coal	155	$0.4	$3
	Biomass	200	$2.6	$13
	Oil	6,424	$45	$7
Facility Energy Audit	All	189,450	$52	$0.3

Source: U.S. EPA, Regulatory Impact Analysis, Table 3-2. EPA did not change these cost estimates as a result of its December 2011 proposed amendments.

Notes: Does not include fuel savings from improving combustion efficiency. Per unit cost was calculated by CRS.

Gas-fired boilers, of which EPA estimates there are 1.3 million, would not be affected by the area source rule. Because the costs of compliance are substantially less than for the MACT rule, the area source rule has not been particularly controversial.

The area source rule would affect approximately 187,000 existing boilers powered by oil, biomass, and coal, located at 92,000 facilities. It would impose annualized costs of $535 million in 2014, according to EPA's Regulatory Impact Analysis. After considering fuel savings from efficiency improvements that would result from the tune-ups required by the rule, the estimated annualized cost is reduced to $487 million.[72] EPA also estimates that about 6,800 new boilers will be constructed at area sources in the next three years; net costs for meeting the area source standards at these facilities are estimated by EPA to be $48 million annually. After accounting for fuel savings from improvements in combustion efficiency, EPA estimates that new sources will experience cost savings of $3.6 million annually rather than incurring compliance costs.[73]

EPA's estimate of costs at area source boilers is summarized in *Table 4.* The table displays EPA data for the promulgated (March 2011) version of the rule. The Fact Sheet accompanying the December 2011 re-proposal states, "We do not expect that the proposed amendments will result in a change in the estimated cost to comply with the rule."[74]

RELATED RULES ON SOLID WASTE INCINERATION

The Boiler MACT and the Area Source Rule were two of four rules related to boilers that EPA promulgated the same day. The other two rules address boilers that use solid waste as fuel and identify what materials EPA considers to be solid waste.

EPA projects that these rules will have less impact than the Boiler MACT, but they address the issues that were at the heart of the court decision that overturned and remanded the boiler rules in 2007.

As noted earlier, the U.S. Court of Appeals for the D.C. Circuit, in *Natural Resources Defense Council v. EPA*, found that EPA had wrongly excluded many industrial boilers from the definition of solid waste incinerators, which have more stringent emissions requirements under Section 129 of the Clean Air Act.[75]

Thus, in addition to the Boiler MACT and Area Source rules, the agency promulgated a rule on the Identification of Non-Hazardous Secondary Materials that Are Solid Waste,[76] and a rule that would set emission standards for Commercial/Industrial Solid Waste Incinerators (the "CISWI Rule"). The first rule identifies solid waste, and the second sets emission standards for the facilities that burn it. On December 2, EPA proposed amendments to both of these rules, as well.[77]

Defining Solid Waste

The purpose of this rule is to clarify which materials are considered solid waste when burned in combustion units and which are not. To be considered solid waste, the basic criterion is whether the material has been discarded. Discarded materials are generally considered solid waste; other materials are not. Whether a material is deemed discarded may be unclear or subject to debate, particularly to a recycler who values the material. EPA addresses this by stating that discarded materials can avoid classification as waste if they meet a number of what it calls "legitimacy criteria":

1) if the material is managed as a valuable commodity;
2) if the material has meaningful heating value (or, for a material considered an ingredient, if it makes a useful contribution to the production or manufacturing process); and
3) if the material contains contaminants at levels comparable to or lower than traditional fuels or ingredients.

Non-hazardous secondary materials that meet legitimacy criteria, such as the following, would not be considered solid waste under the rule promulgated in March 2011:

- material used as a fuel that remains within the control of the generator (whether at the site of generation or another site the generator has control over);
- scrap tires removed from vehicles and managed under established tire collection programs;
- resinated wood residuals, provided they have not been discarded and are used as fuel either by the generator or outside the generator's control;
- material used as an ingredient in a manufacturing process (whether by the generator or a third party);
- material that has been sufficiently processed to produce a fuel or ingredient product; and
- material that has been determined through a case-by-case petition process to not have been discarded and to be indistinguishable in all relevant aspects from a fuel product.[78]

Controversy over this rule centered on how EPA would interpret these criteria for certain recovered materials that are commonly used as fuel, particularly "off-spec" used oil and whole scrap tires. The originally proposed rule did not specifically identify these materials as solid waste. However, in the preamble to the proposal, EPA did identify these materials as solid waste even when they are used as fuel.

Used Oil

EPA defines used oil as either complying with limits for contaminants of concern ("on-spec") or not ("off-spec"). On-spec oil is exempt from waste management regulations, because the contaminants in it are either at the same concentration or at a lower concentration than in virgin refined fuel oil. Off-spec used oil, on the other hand, even if it is managed within the control of the generator, contains contaminants at levels that are not comparable to traditional fuels, and thus would not be considered a non-waste fuel under the legitimacy criteria described above.

Under previous regulations promulgated under the Resource Conservation and Recovery Act (RCRA, 40 CFR part 279), off-spec used oil could be burned in used-oil-fired space heaters, provided that, in EPA's words:

> (1) The heater burns only used oil that the owner or operator generates or used oil received from household do-it-yourself used oil generators; (2) the heater is designed to have a maximum capacity of not more than 0.5 million Btu per hour; and (3) the combustion gases from the heater are vented to the ambient air. The RCRA used oil regulations base this provision on a finding that uncontrolled emissions from these sources do not pose a significant threat to human health and the environment. (Used Oil Final Rule, 50 FR 49194 (November 29, 1985).) However, consistent with our determination that off-spec used oil be considered a solid waste when burned as a fuel, we believe that off-spec used oil managed within the control of the generator would not qualify for the generator controlled exclusion when burned in a used oil fired-space heater, since contaminant levels are not comparable to traditional fuels. Therefore, we are proposing that off-spec used oil combusted at a unit that is within the control of the generator would be solid waste.79

If the used oil were classified as solid waste, then the space heater would have to meet the "CISWI" incinerator standards described below, which no space heater is likely to meet.

Most used oil is considered on-spec, but many of those who commented on the proposal argued that unless there is a general exclusion such as that written into the existing RCRA regulations, it would be necessary to test the oil and determine that it is on-spec before burning it in a space heater. Doing so would be costly and impractical. Thus, the proposed regulations, in the minds of many commenters, would have had the practical effect of banning the use of waste oil in space heaters.

The proposal also appeared to contradict the existing RCRA regulations, but did not specifically repeal them. As a result, the Code of Federal Regulations might have contained two conflicting sets of rules applicable to the combustion of used oil.

The final rule promulgated in March 2011 clarified these issues:

> EPA is specifically clarifying in this final rule that used oil combusted in an oil-fired space heater that meets the provisions of 40 CFR 279.23 [i.e., the existing RCRA regulations] need not be tested to establish whether or not such oil is on or off-spec. This includes used oil generated by small facilities such as auto repair shops and machine shops that have such units, and used oil-generated by homeowners who change their own oil (referred to as "do-it-yourself" or "DIY" oil) that are burned in such units. This is because the CISWI regulations promulgated elsewhere in the Federal Register today do not establish emissions limits for such units, and therefore the concerns of the commenters that such units would have to comply with CAA Section 129 standards have been addressed for this population of combustion units.80

Scrap Tires

The rule proposed in June 2010 also would have imposed new restrictions on the use of scrap tires as fuel:

> whole used tires (even if collected from tire dealerships and automotive shops and overseen by a state tire collection oversight program) are initially abandoned and thus meet the plain meaning of discard. As a result, whole used tires that are not processed into a legitimate fuel or ingredient (e.g., shredded/chipped with steel belts removed) would be considered a solid waste. We acknowledge that whole tires can be legitimately burned as fuel, but because they have been discarded, whole tires would be considered solid wastes and subject to the CAA section 129 requirements unless processed into a non-waste fuel product.81

This would have been a change from current policy and would have affected the use of scrap tires as fuel.

EPA reversed itself in the March 2011 final rule:

> After careful consideration of the comments and all the material in the rulemaking record, including documents cited in the ANPRM [Advance Notice of Proposed Rulemaking] and the preamble to the proposed rule, the Agency agrees that a system where scrap tires are removed from vehicles and are collected and managed under the oversight of established tire collection programs are not "discarded in the first instance." Such tires (including both whole tires and tires that have been shredded—with or without metal removal) are non-waste when used as a fuel in combustion units. These programs ensure that the tires are not discarded en route to the combustor for use as fuel and are handled as a valuable commodity as required in the legitimacy criterion in today's rule at § 241.3(d)(1)(i).82

Further Clarification

In the December 2 re-proposal, EPA has attempted to further clarify what fuels would be considered non-hazardous secondary materials (NHSM), as opposed to solid waste, generally broadening the definitions and criteria to permit more such materials to be used in boilers without the boiler being considered an incinerator.[83]

The CISWI Rule

The Commercial/Industrial Solid Waste Incinerator (CISWI) rule promulgated on March 21 would set emission standards for commercial and industrial facilities that burn materials determined to be solid waste (i.e., materials that do not meet the above criteria).[84] CISWI's emission standards are required to be set under Section 129 of the Clean Air Act, which has more stringent requirements than Section 112. In addition to the five groups of pollutants addressed by the Boiler MACT, the CISWI rule sets emission limits for lead, cadmium, sulfur dioxide, and nitrogen oxides. Section 129 also makes no distinction between major sources and area sources, thus setting the more stringent MACT standards for smaller facilities.

EPA has identified five subcategories of CISWI facilities: incinerators, energy recovery units for solids, energy recovery units for liquids and gases, waste burning kilns, and small remote incinerators—a total of 95 existing facilities. The total nationwide annualized costs of compliance for these units were estimated to be $270 million. EPA estimates the benefits of the final rule, including the avoidance of 40-100 premature deaths annually, at $330 million-$800 million.[85]

Controversy over the originally proposed CISWI standards focused on a category called "burn-off ovens." Burn-off ovens, as defined by EPA in the proposed rule, are units that combust residual materials off racks, parts, drums or hooks so that those items can be re-used in various production processes. Operators of such facilities stated that there are more than 15,000 such units (EPA had identified 36), and they maintained that the units should not be characterized as incinerators, but should be considered boilers, subject to either the Boiler MACT or the Area Source rule. In the final CISWI rule, EPA concluded that it didn't have sufficient data for burn-off ovens, and removed them and several other types of units from the incinerator definition. The preamble to the December 2 proposal repeats the agency's conclusion that it does not have data that would allow it to establish standards for the various burn-off oven subcategories.

CONCLUSION

EPA's Boiler MACT remains controversial. The version of the rule promulgated March 21, 2011, and the amendments proposed December 2, 2011, are much less stringent than the rule as first proposed, and the standards could change further before the reconsideration process now underway is completed. The proposed amendments to the rules have been subject to further public comment and the agency hopes to finalize them by late spring 2012.

Members of Congress have been active participants in EPA's public comment process (more than 100 Members of the House and more than 40 Senators wrote EPA regarding the originally proposed rule) and they remain interested in these rulemakings. Bills have been introduced in both the House and Senate (H.R. 2250 and S. 1392) to change the Clean Air Act requirements for these rules and the deadlines for implementation. The bills would provide additional time for implementation of standards and would change key aspects of the Section 112 requirements as they apply to boilers and CISWI units. They would revoke the standards promulgated on March 21, 2011, and require promulgation of replacements for the Boiler MACT and related rules 15 months after the date of the bills' enactment; and they would require EPA to set a compliance date no earlier than five years after the date of promulgation. At a minimum, this would give the affected units three years of additional time to comply with MACT standards.

The bills would also make substantive changes in the Section 112 requirements as applied to boilers and CISWI units. They would remove the requirements that currently apply in the absence of EPA regulation—what are called the "MACT hammers": under current law, permits issued in the absence of MACT regulations are required to include MACT emission limits determined on a case-by-case basis. The bills define certain sources currently considered as "new" to be "existing" sources, which would be subject to less stringent requirements. And they set less stringent requirements for the standards themselves, requiring EPA to choose the "least burdensome" regulatory alternative, and requiring that standards can be met "consistently and concurrently with emission standards for all other air pollutants," which might prohibit EPA's use of the "pollutant by pollutant" approach that it used in setting the currently promulgated standards.

H.R. 2250 passed the House, 275-142, on October 13. A Senate version similar to H.R. 2250 was offered as an amendment (S.Amdt. 1660) to the surface transportation bill (S. 1813) on March 8, 2012, but was not adopted on a vote of 52-46 (60 votes being necessary for adoption).

End Notes

[1] The proposed changes appeared in the *Federal Register* December 23, 2011 (76 *Federal Register* 80598). The *Federal Register* notice and background information are available at http://www.epa.gov/airquality/combustion/ actions.html. The March 2011 standards are at U.S. EPA, "National Emission Standards for Hazardous Air Pollutants for Major Sources: Industrial, Commercial, and Institutional Boilers and Process Heaters; Final Rule," 76 *Federal Register* 15608, March 21, 2011, which is also accessible through EPA's boiler webpage.

[2] United States Sugar Corp v. EPA, No. 11-1108 (D.C. Cir. filed April 14, 2011).

[3] The rules, as promulgated, had effective dates of May 20, 2011, although compliance with their emission standards would not have been required until at least three years later.

[4] Sierra Club v. Jackson, No. 11-278 (D.D.C. Jan. 9, 2012).

[5] The No Action Assurance Letter, dated February 7, 2012, can be found at http://www. epa.gov/ ttn/atw/boiler/ boiler_ciswi-no_action_2012-02-07.pdf.

[6] 76 *Federal Register* 80616, December 23, 2011.

[7] Letter of EPA Administrator Lisa P. Jackson to Senator Ron Wyden, March 5, 2012, at http://www. eenews.net/assets/ 2012/03/05/document_pm_02.pdf.

[8] U.S. EPA, "National Emission Standards for Hazardous Air Pollutants; Notice of Reconsideration," 76 *Federal Register* 15266, March 21, 2011.

[9] The May 16 announcement appeared in the May 18 *Federal Register*. See http://www.gpo. gov/fdsys/pkg/FR-2011- 05-18/pdf/2011-12308.pdf.

[10] The data in this paragraph are from U.S. EPA, Fact Sheet, "Proposed Amendments to the Air Toxics Standards for Industrial, Commercial, and Institutional Boilers at *Major* Source

Facilities," December 2011, at http://www.epa.gov/ airquality/combustion/docs/20111202 msboilerfs.pdf, and the *Federal Register* notice, at http://www.gpo.gov/fdsys/ pkg/FR-2011-12-23/pdf/2011-31667.pdf.

[11] U.S. EPA, "National Emission Standards for Hazardous Air Pollutants for Major Sources: Industrial, Commercial, and Institutional Boilers and Process Heaters; Proposed Rule," 75 *Federal Register* 32048, June 4, 2010. Also, see Comments of Clean Air Task Force, Earthjustice, Natural Resources Defense Council, and the Sierra Club on National Emission Standards for Major Sources: Industrial, Commercial, and Institutional Boilers and Process Heaters: Proposed Rule, Docket ID No. EPA-HQ-OAR-2002-0058, August 23, 2010, p. 3.

[12] U.S. EPA, "National Emission Standards for Hazardous Air Pollutants for Major Sources: Industrial, Commercial, and Institutional Boilers and Process Heaters," pre-publication copy, December 2, 2011, p. 141, at http://www.epa.gov/ airquality/combustion/docs/ 20111202msboilerproposal.pdf. Hereinafter referred to as "Proposed Amendments."

[13] Natural Resources Defense Council v. EPA, 489 F. 3d 1250, 1257–61 (DC Cir. 2007).

[14] Sierra Club v. Jackson, No. 01-1537, 2001 Westlaw 181097, at *7 (D.D.C. January 20, 2011).

[15] Sierra Club v. Johnson, 444 F. Supp. 2d 46, 47 (D.D.C. 2006). This case is separate from the vacatur and remand of the Boiler MACT in the 2007 NRDC v. EPA decision. In effect, there are two courts in two separate proceedings overseeing EPA's boiler rules.

[16] *Sierra Club*, 2001 Westlaw 181097.

[17] U.S. EPA, "EPA Announces Next Steps on Emissions Standards for Boilers, Certain Incinerators," Press Release, January 20, 2011, at http://yosemite.epa.gov/opa/ admpress. nsf/6424ac1caa800aab85257359003f5337/ 58f5bee5e13c61228525781e007e9881!OpenDocument.

[18] The other two subcategories were only required to meet work practice standards. Work practice standards require certain actions from the regulated entity (e.g., the boiler must be tuned up annually, or the owner must conduct an assessment to identify energy conservation opportunities), but they don't set numeric standards for emissions.

[19] The three types of coal-fired boiler identified in the proposed rule were coal stoker, coal fluidized bed, and pulverized coal. The four types of biomass-fired boilers were biomass stoker, biomass fluidized bed, biomass suspension burner/Dutch oven, and biomass fuel cells. In addition, the agency proposed emission limits for liquid-fueled boilers, and gas-fired boilers using "other process gases."

[20] Hydrogen chloride is often referred to as hydrochloric acid, because when the gas encounters water in the atmosphere it forms an acidic solution of hydrochloric acid. The specific emission limits EPA proposed for each of the five pollutants can be found in the June 4, 2010, *Federal Register* at p. 32012, Table 1.The final standards are at 76 *Federal Register* 15612, Table 1, March 21, 2011.

[21] EPA has acknowledged that it did not have as much data as it might have wished to use in establishing the boiler subcategories and the proposed MACT standards. In a September 28, 2010, letter, the Administrator stated: "In an effort to establish subcategories wherever appropriate, and to calculate accurately the standards for each subcategory, EPA asked the affected companies and institutions for technical data about their facilities long before the court-ordered deadline for publishing a proposal. As is often the case in §112 rulemaking efforts, however, EPA did not receive many data. While the agency was not left entirely lacking in relevant information, the limited response from affected businesses and institutions did make it difficult for EPA to delineate subcategories and calculate standards that fully reflected operational reality. The agency nevertheless was legally required to publish proposed subcategories and standards based on the information it had at the time."

Letter of EPA Administrator Lisa P. Jackson to Senator Mary L. Landrieu, September 28, 2010, p. 2.

[22] The agency defined limited use units as those having a federally enforceable limit of no more than 876 hours per year of operation (i.e., operable at most 10% of the year). 76 *Federal Register* 15684, March 21, 2011.

[23] In the rule's preamble, EPA stated that there was a "need for a noncontinental liquid fuel subcategory for island units that have limited fuel options and other unique circumstances." 76 *Federal Register* 15633, March 21, 2011. See also discussion at p. 15635.

[24] Medical Waste Institute et al. v. EPA, No. 09–1297 (D.C. Cir.).

[25] U.S. EPA, "Standards of Performance for New Stationary Sources and Emissions Guidelines for Existing Sources: Hospital/Medical/Infectious Waste Incinerators; Final Rule," October 6, 2009, 74 *Federal Register* 51381.

[26] Ibid.

[27] Ibid. The case in question was Sierra Club v. EPA, 167 F.3d 658, 660 (D.C.Cir. 1999).

[28] The full discussion is found in Section V.A. of the Preamble, at 76 *Federal Register* 15621-15623, March 21, 2011.

[29] 76 *Federal Register* 15612, Table 1, March 21, 2011.

[30] Letter of EPA Administrator Lisa P. Jackson to Senator Mary L. Landrieu, September 28, 2010, p. 2.

[31] Proposed Amendments to the Boiler MACT Rule, pre-publication copy, December 2, 2011, p. 136.

[32] A rough rule of thumb for coal is that it contains about 10,000 Btus of energy per pound. To be more precise, the heating value ranges from 6,500 to 13,000 Btus per pound, depending on rank (i.e., type of coal), with bituminous coal containing more than 10,000 and subbituminous and lignite less.

[33] IHS Global Insight for CIBO, *The Economic Impact of Proposed EPA Boiler/Process Heater MACT Rule on Industrial, Commercial, and Institutional Boiler and Process Heater Operators*, August 2010, Appendix A, p. 28.

[34] Bill Becker, "Clean Air Issues Facing States and Localities: Regulatory Update," presentation, Institute of Clean Air Companies, April 27, 2011.

[35] For a further discussion of variability analysis, see Amanda Singleton, ERG (Eastern Research Group, Inc.), "Revised MACT Floor Analysis (2011) for the Industrial, Commercial, and Institutional Boilers and Process Heaters National Emission Standards for Hazardous Air Pollutants – Major Source," Memorandum to Brian Shrager, U.S. EPA, January 4, 2011, at http://www.epa.gov/ttn/atw/boiler/rev_mact_floor_analysis_major_boilers_process_heaters.pdf.

[36] Proposed Amendments to the Boiler MACT Rule, pre-publication copy, December 2, 2011, pp. 134-135.

[37] U.S. EPA, Proposed Boiler MACT, 75 *Federal Register* 32019, June 4, 2010.

[38] Ibid.

[39] Ibid.

[40] See, for example, the U.S. Energy Information Administration's *Annual Energy Outlook 2010 with Projections to 2035*, at http://www.eia.doe.gov/oiaf/aeo/gas.html: "A 4-fold increase in shale gas production from 2008 to 2035 more than offsets a 31-percent decline in other lower 48 onshore natural gas production in the AEO2010 Reference case. Significant increases in shale gas production are expected in the Northeast, Gulf Coast, and Midcontinent regions...." Also, CRS Report R41543, *Global Natural Gas: A Growing Resource*, by Michael Ratner.

[41] U.S. EPA, Regulatory Impact Analysis, Appendices A and B, February 23, 2011, at http://www.epa.gov/ttn/ecas/ regdata/RIAs/boilersriafinal110221_psg.pdf. Hereafter referred to as "Regulatory Impact Analysis."

[42] EPA Fact Sheet, p. 2, at http://www.epa.gov/airquality/combustion/ docs/20111202msboilerfs.pdf.

[43] EPA National Emissions Inventory, "Trends in Mercury Air Emissions Between 1990 and 2005," June 28, 2010, and U.S. EPA, "Emissions Overview: Hazardous Air Pollutants in Support of the Proposed Toxics Rule," March 15, 2011, available on request.

[44] Personal communication, U.S. EPA, Office of Air Quality Planning and Standards, December 15, 2011.

[45] Proposed Amendments, previously cited, p. 141.

[46] Regulatory Impact Analysis, previously cited, p. 7-3. In the qualitative discussion of the benefits of reducing HAP emissions, the RIA states that the effects of exposure to HAP emissions can include neurological, cardiovascular, liver, kidney, and respiratory effects as well as effects on the immune and reproductive systems and cancer. Reducing emissions may reduce these effects, but none of these benefits were quantified in the benefit estimates. As summarized by the agency's Science Advisory Board, "(T)he challenges for assessing progress in health improvement as a result of reductions in emissions of hazardous air pollutants (HAPs) are daunting ... due to a lack of exposure-response functions, uncertainties in emissions inventories and background levels, the difficulty of extrapolating risk estimates to low doses and the challenges of tracking health progress for diseases, such as cancer, that have long latency periods." (RIA, p. 7-43) As a result, the agency concluded, "Large reductions in HAP emissions may not necessarily translate into significant reductions in health risk because toxicity varies by pollutant and whether or not there are exposures at or above levels of concern is not known." (RIA, p. 7-41)

[47] IHS Global Insight for CIBO, *The Economic Impact of Proposed EPA Boiler/Process Heater MACT Rule on Industrial, Commercial, and Institutional Boiler and Process Heater Operators*, August 2010, pp. 29-30. Hereinafter, "CIBO Study."

[48] CIBO Study, pp. 9-10. The "at risk" jobs include direct impacts on jobs at the facilities that have to install pollution controls, indirect impacts on jobs at suppliers of those firms, and induced impacts due to reductions in spending by employees in the direct and indirect categories.

[49] Draft Regulatory Impact Analysis, pp. 1-2 and 4-6 to 4-9.

[50] Proposed Amendments, previously cited, p. 138.

[51] CIBO study, p. 11.

[52] Letter of EPA Administrator Lisa P. Jackson to Senator Mary L. Landrieu, September 28, 2010, p. 3. Others have also discussed the CIBO study's failure to account for the increased output in the pollution control industry, along with other criticisms of the study. See, for example, Laurie Johnson, "Boiler Industry's Junk Economic Analysis of Proposed EPA Toxic Emission Standards for Industrial Boilers," NRDC Switchboard, http://switchboard.nrdc.org/ blogs/ljohnson/boiler_industrys_junk_analysis.html.

[53] National Association of Clean Air Agencies, *EPA's Proposed Regulations on Hazardous Air Pollutants from Boilers*, December 8, 2010, at http://www.4cleanair.org/Documents/ EVALUATIONOFCIBOSTUDYFINAL12082010.pdf. Hereinafter, the "NACAA study."

[54] NACAA study, p. 5.

[55] Ibid.

[56] Ibid., pp. 9-10.

[57] Fisher International Inc. for the American Forest and Paper Association, "Economic Impact of Pending Air Regulations on the U.S. Pulp and Paper Industry," August 2010, 8 p. The revised report is entitled, "Economic Impact of Pending Air Regulations on the U.S. Pulp and Paper Industry - Updated," September 2011, 17 p.

[58] The study also estimated capital costs of $950 million for compliance with the CISWI rule, discussed separately below. Because the earlier study lumped the costs of the two rules together, the remainder of this discussion does the same (i.e., our reference to AF&PA's revised cost estimates refers to the combined cost of both rules).

[59] The closest it comes to an explanation is a statement that average costs of production in U.S. mills increased between the two study periods and "higher costs might put more U.S. mills at risk rather than fewer because foreign mills gain a cost advantage vis-à-vis U.S. mills." AF&PA Updated Study, p. 8.

[60] Administrator Lisa P. Jackson, "Remarks on the 40th Anniversary of the Clean Air Act, As Prepared," September 14, 2010, at http://yosemite.epa.gov/opa/admpress.nsf/ 8d49f7ad4bbc f4ef852573590040b7f6/ 7769a6b1f0a5bc9a8525779e005ade13!OpenDocument.

[61] NACAA study, p. 4.

[62] Letter of EPA Administrator Lisa P. Jackson to Senator Mary L. Landrieu, September 28, 2010, p. 3.

[63] Comments of Clean Air Task Force, Earthjustice, Natural Resources Defense Council, and the Sierra Club on National Emission Standards for Major Sources: Industrial, Commercial, and Institutional Boilers and Process Heaters: Proposed Rule, Docket ID No. EPA-HQ-OAR-2002-0058, August 23, 2010, p. 3.

[64] Ibid.

[65] NACAA study, pp. 5-8.

[66] Boiler MACT proposal, 75 *Federal Register* 32011, June 4, 2010.

[67] For more information on the 112(d)(4) issue, see the discussion in the Boiler MACT proposal at 75 *Federal Register* 32030-32033, June 4, 2010.

[68] Letter of Senator Mary L. Landrieu at al. to EPA Administrator Lisa Jackson, September 24, 2010, p. 2.

[69] The area source rule, as promulgated, is at 76 *Federal Register* 15554, March 21, 2011. The December 2011 re-proposal is at http://www.epa.gov/airquality/combustion/docs/ 20111202 asproposal.pdf.

[70] U.S. EPA, "Final Air Toxics Standards for Industrial, Commercial, and Institutional Boilers at Area Source Facilities," Fact Sheet, p. 2, at http://www.epa.gov/airquality/combustion/docs/ 20110221aboilersfs.pdf.

[71] Ibid., p. 5. The actual standards can be found at 76 *Federal Register* 15559, Table 1. These standards do not address most of the pollutants covered by the major source MACT. Compared to the major source MACT, they are also less stringent for the pollutants that they do address.

[72] Regulatory Impact Analysis, previously cited, p. 3-6.

[73] 76 *Federal Register* 15579, March 21, 2011.

[74] Area Source Fact Sheet, December 2011, p. 3, at http://www.epa.gov/airquality/ combustion/ docs/ 20111202asboilersfs.pdf.

[75] Natural Resources Defense Council v. EPA, 489 F. 3d 1250, 1257–61 (DC Cir. 2007).

[76] 76 *Federal Register* 15456, March 21, 2011.

[77] These rules also appeared in the December 23, 2011, *Federal Register*, at 76 *Federal Register* 80452.

[78] Ibid., pp. 15459-15460.

[79] U.S. EPA, "Identification of Non-Hazardous Secondary Materials that Are Solid Waste; Proposed Rule," 75 *Federal Register* 31865, June 4, 2010.

[80] 76 *Federal Register* 15502, March 21, 2011.

[81] U.S. EPA, "Identification of Non-Hazardous Secondary Materials that Are Solid Waste; Proposed Rule," 75 *Federal Register* 31864, June 4, 2010.

[82] 76 *Federal Register* 15491-15492, March 21, 2010 [footnotes omitted].

[83] The agency identified numerous materials that it considers to be within the definition of "clean cellulosic biomass," and thus are non-waste fuels, including "agricultural derived biomass, other crop residues (including vines, orchard trees, hulls, seeds), other biomass crops used for the production of cellulosic biofuels, hogged fuel, untreated wood pallets, wood pellets, and wood debris from urban areas." The agency also identified two secondary materials, resinated wood products and tires managed under the oversight of established tire collection programs, as non-wastes, stating further that it "recognizes that certain NHSMs may not meet the legitimacy criteria in all instances, but after balancing the legitimacy criteria with other relevant factors, the material would still generally be considered a non-waste fuel." In addition, the agency is seeking comment on whether pulp and paper wastewater treatment sludges and coal refuse from legacy piles should also be categorically identified as non-wastes. Finally, EPA proposes to revise the legitimacy criteria to clarify that for the purposes of deciding whether a material contains contaminants at levels comparable to or lower than traditional fuels or ingredients, NHSMs may be compared to any traditional fuel that a combustion unit is designed to burn; also, comparisons can be made for groups of pollutants (such as volatile organic compounds) rather than individual contaminants. For more information, see U.S. EPA, "Fact Sheet: Proposed Revisions to the Identification of Non-Hazardous Secondary Materials That Are Solid Wastes Final Rule," at http://www.epa.gov/wastes/nonhaz/ define/pdfs/fs12-2-11.pdf.

[84] U.S. EPA, "Standards of Performance for New Stationary Sources and Emission Guidelines for Existing Sources: Commercial and Industrial Solid Waste Incineration Units; Final Rule," 76 *Federal Register* 15704, March 21, 2011.

[85] The source of the mortality data was U.S. EPA, "Final Amendments to New Source Performance Standards and Emission Guidelines for Commercial and Industrial Solid Waste Incineration Units," Fact Sheet, February 21, 2011, pp. 2-3, at http://www.epa.gov/ airquality/combustion/docs/20110221ciswifs.pdf. The agency states in the prepublication version of the December 2 amendments that, "Due to last minute changes to the March 2011 final CISWI rule, we were unable to incorporate the final engineering costs and emission reductions into the RIA, which would decrease the costs by approximately 22 percent and increase the monetized benefits by approximately 4 percent from those shown here." See Proposed Amendments to the CISWI Rule, p. 79, at http://www.epa.gov/ airquality/combustion/ docs/20111202ciswiproposal.pdf.

In: Federal Control of Air Emissions
Editors: H.M. Graham and T.E. Lee
ISBN: 978-1-62081-808-4

Chapter 3

AIR QUALITY: INFORMATION ON TALL SMOKESTACKS AND THEIR CONTRIBUTION TO INTERSTATE TRANSPORT OF AIR POLLUTION*

United States Government Accountability Office

ABBREVIATIONS

CAIR	Clean Air Interstate Rule
CAMx	Comprehensive Air Quality Model with Extensions
DOE	Department of Energy
EIA	Energy Information Administration
EPA	Environmental Protection Agency
ESP	electrostatic precipitator
FGD	flue gas desulfurization
GEP	good engineering practice
NAAQS	National Ambient Air Quality Standards
NESCAUM	Northeast States for Coordinated Air Use Management
NO_x	nitrogen oxides
SCR	selective catalytic reduction

* This is an edited, reformatted and augmented version of The United States Government Accountability Office publication, GAO-11-473, dated May 2011.

SNCR	selective non-catalytic reduction
SIP	State Implementation Plan
SO_2	sulfur dioxide

WHY GAO DID THIS STUDY

Tall smokestacks—stacks of 500 feet or higher, which are primarily used at coal power plants—release air pollutants such as sulfur dioxide (SO_2) and nitrogen oxides (NO_x) high into the atmosphere to help limit the impact of these emissions on local air quality. Tall stacks can also increase the distance these pollutants travel in the atmosphere and harm air quality and the environment in downwind communities. The 1977 amendments to the Clean Air Act encourage the use of pollution control equipment over dispersion techniques, such as tall stacks, to meet national air standards. Section 123 of the Act does not limit stack height, but prohibits sources of emissions from using the dispersion effects of stack heights in excess of a stack's good engineering practice (GEP) height to meet emissions limitations.

GAO was asked to report on (1) the number and location of tall stacks of 500 feet or higher at coal power plants and when they began operating; (2) what is known about such stacks' contribution to the interstate transport of air pollution and the pollution controls installed at plants with these stacks; and (3) the number of stacks that were built above GEP height since 1988 and the reasons for this. GAO analyzed Energy Information Administration (EIA) data on power plants, surveyed states with tall stacks, and interviewed experts on the transport of air pollution. GAO is not making recommendations in this report. The Environmental Protection Agency and the Department of Energy stated they had no comments on this report.

WHAT GAO FOUND

According to analysis of EIA data, which were updated with GAO's survey results, a total of 284 tall smokestacks were operating at 172 coal power plants in 34 states, as of December 31, 2010. Of these stacks, 207 are 500 to 699 feet tall, 63 are 700 to 999 feet tall, and the remaining 14 are 1,000 feet tall or higher. About one-third of these stacks are concentrated in 5 states along the Ohio River Valley. While about half of tall stacks began operating more than 30 years ago, there has been an increase in the number of tall stacks that began operating in the last 4 years, which air and utility officials attributed to the need for new stacks when plants installed pollution control equipment.

Stack height is one of several factors that contribute to the interstate transport of air pollution. According to reports and stakeholders with expertise on this topic, tall stacks generally disperse pollutants over greater distances than shorter stacks and provide pollutants greater time to react in the atmosphere to form ozone and particulate matter. However, stakeholders had difficulty isolating the exact contribution of stack height to the transport of air pollution because this is a complex process that involves several other variables, including total emissions from a stack, the temperature and velocity of the emissions, and weather. The use of pollution controls, which are generally installed in boilers or the duct work that connects a boiler to a stack, has increased in recent years at coal power plants. However, GAO found that many boilers remain uncontrolled for certain pollutants, including several connected to tall stacks. For example, GAO found that 56 percent of boilers attached to tall stacks lacked scrubbers to control SO_2 and 63 percent lacked post-combustion controls to capture NO_x emissions. In general, GAO found that boilers without these controls tended to be older, with in-service dates prior to 1980.

GAO identified 48 tall stacks built since 1988—when GEP regulations were largely affirmed in court—that states reported are subject to the GEP provisions of the Clean Air Act and for which states could provide GEP height information.

Of these 48 stacks, 17 exceed their GEP height, 19 are at their GEP height, and 12 are below their GEP height. Section 123 of the Clean Air Act defines GEP as the height needed to prevent excessive downwash, a phenomenon that occurs when nearby structures disrupt airflow and produce high local concentrations of pollutants. Officials reported that a variety of factors can influence stack height decisions. For example, some utility officials reported that stacks were built above GEP to provide greater protection against downwash or to help a plant's emissions clear local geographic features, such as valley walls.

GAO was unable to obtain GEP height information for an additional 25 stacks that were built since 1988 for two reasons: (1) some of these stacks were exempt from GEP regulations, and (2) states did not have GEP information readily available for some replacement stacks because the GEP calculation was sometimes made decades earlier and a recalculation was not required at the time the replacement stack was built.

May 11, 2011

The Honorable Sheldon Whitehouse
Chairman
Subcommittee on Oversight
Committee on Environment and Public Works United States Senate

Dear Mr. Chairman:

Tall smokestacks—which are used primarily at coal power plants—release air pollutants such as sulfur dioxide (SO_2), and nitrogen oxides (NO_x) high into the atmosphere to help disperse them and limit their impact on air quality in local communities.[1] However, because wind currents are generally faster at higher elevations, tall stacks can increase the distance that these pollutants travel, harming air quality in downwind communities. When these pollutants are airborne, they can react in the atmosphere to form particulate matter, acid rain, and ozone that can harm air quality, human health, and the environment. For example, SO_2, NO_x, ozone, and particulate matter can cause or worsen respiratory discascs such as emphysema, bronchitis, or asthma, while acid rain can damage vegetation and aquatic ecosystems.

Under the Clean Air Act, the Environmental Protection Agency (EPA) is responsible for setting National Ambient Air Quality Standards (NAAQS) for certain pollutants considered harmful to public health and the environment. EPA has set NAAQS for six such pollutants, known as criteria air pollutants: SO_2, NO_x, particulate matter, ozone, carbon monoxide, and lead. These standards are expressed as concentration limits averaged over time, and compliance is determined through ground-level monitoring at a local level. States are responsible for developing and implementing plans, known as State Implementation Plans (SIP), to achieve and maintain these standards. In carrying out this duty, states set emissions limitations for individual sources of air pollution, which are based, in part, on the results of air quality models that show the impact these sources will have on air quality. Since 1990, the Clean Air Act has required the incorporation of these emissions limitations into operating permits which collect all of the pollution control, recordkeeping, and reporting requirements applicable to individual sources of air pollution.

In the early 1970s, power plants commonly installed tall stacks to reduce pollutant concentrations at ground level to help attain NAAQS. The 1977 amendments to the Clean Air Act encouraged the use of pollution control equipment and other control measures over dispersion techniques such as tall stacks to meet NAAQS. For example, section 123 was added to prohibit states from counting the dispersion effects of stack heights in excess of a stack's good engineering practice (GEP) height when determining a source's emissions limitation. Section 123 of the Clean Air Act defines GEP height as the height needed to disperse pollutants to prevent excessive "downwash," a phenomenon that occurs when nearby structures disrupt airflow and produce excessively high concentrations of pollutants in the immediate vicinity of the source. Section 123 generally applies to stacks built since December 31, 1970, but some stacks may be exempt if they were built to replace stacks that were in existence on or before this date. Since the GEP heights for smokestacks can be determined using a calculation that accounts for the height and width of the largest nearby structure, GEP heights vary accordingly.[2] Section 123 does not limit stack height; instead, it removes an incentive to build stacks higher than necessary. For example, if a stack's GEP height is 600 feet, but the stack is built to 800 feet, the source cannot count the dispersion effects associated with the excess 200 feet toward meeting its emissions limitation. EPA finalized regulations for calculating and using GEP height in 1985, and these regulations were largely affirmed by the District of Columbia Court of Appeals in 1988.

EPA reported that measured levels of SO_2 and NO_x, along with ozone and particulate matter, decreased between 1990 and 2008. However, EPA noted that in 2008, about 127 million people lived in counties where one or more NAAQS—usually ozone or particulate matter—was exceeded. In developing policy to control air pollution, EPA recognizes that emissions from upwind states can contribute to the nonattainment—or exceedances—of NAAQS in downwind states. EPA has taken steps to reduce SO_2 and NO_x emissions that contribute to the interstate transport of air pollution through recent rule makings. You asked us to provide information on the use of tall smokestacks at coal power plants. Specifically, our objectives were to examine (1) the number and location of smokestacks 500 feet or higher that are operating at coal power plants across the United States, and when they began operating; (2) what is known about these smokestacks' contribution to the interstate transport of air pollution and the pollution controls that have been installed at coal power plants with these stacks; and (3) the number of these smokestacks that were built above their GEP height since 1988, and the reasons for this.

To identify the number and location of smokestacks at coal power plants that were 500 feet or higher on December 31, 2010, we analyzed data on power plants from the Department of Energy's (DOE) Energy Information Administration (EIA). We also used these data to determine when these stacks began operating. To assess the reliability of the EIA data used in this report, we reviewed documentation from EIA, interviewed relevant officials who were involved in collecting and compiling the data, and conducted electronic testing of the data. We determined that the data were sufficiently reliable for our purposes. Because the EIA data were collected in 2008, we also contacted all 50 states and the District of Columbia and sent a survey to states with tall stacks to determine if any changes had taken place in the number or operating status of stacks since that time. We updated the relevant EIA data with more recent data from our survey results. To determine what is known about tall stacks' contribution to the interstate transport of air pollution, we reviewed reports from EPA and academics and spoke with stakeholders with expertise on this topic. These stakeholders included EPA officials involved in modeling interstate transport of air pollution from power plants, officials from utilities and construction firms that design and build power plants, atmospheric scientists who conduct research on this topic, and state officials who are involved in permitting power plants and complying with federal regulations governing the interstate transport of air pollution. We also analyzed the EIA data to determine the pollution control equipment installed at coal power plants with stacks 500 feet or higher. To determine the number of tall stacks that have been built above their GEP height since 1988, we used survey responses from 22 states in which tall stacks have been built since 1988 to obtain information about the GEP height for these stacks. In this survey, we also asked for reasons that a stack was built above GEP, when applicable. In those cases where state officials could not provide a reason for why a stack was built above its GEP height, we contacted several of the operators of these facilities to obtain this information. We conducted our work from July 2010 through May 2011 in accordance with all sections of GAO's quality assurance framework that are relevant to our objectives. This framework requires that we plan and perform the engagement to obtain sufficient, appropriate evidence to meet our stated objectives and to discuss any limitations in our work. We believe that the information and data obtained, and the analysis conducted, provide a reasonable basis for any findings and conclusions. A more detailed description of our scope and methodology is presented in appendix I.

BACKGROUND

The five principal emissions from coal power plants are carbon dioxide, SO_2, NO_x, particulate matter, and mercury. For the purposes of this report, we are focusing on power plants' emissions of SO_2, NO_x, and particulate matter since they, along with ozone, are the focus of a rule currently proposed by EPA—the Transport Rule—which seeks to limit the interstate transport of emissions of SO_2 and NO_x in order to abate violations of particulate matter and ozone NAAQS in downwind states. According to an EPA analysis, as of 2008, power plants emitted over 65 percent of SO_2 emissions and almost 20 percent of NO_x emissions, nationwide. These emissions impact local air quality, but they can also travel hundreds of miles to impact the air quality of downwind states. In developing the Transport Rule, EPA has found that emissions of SO_2 and NO_x from 31 eastern states and the District of Columbia prevent downwind states from meeting NAAQS for ozone and particulate matter. SO_2 and NO_x emissions contribute to the formation of fine particulate matter, and NO_x emissions contribute to the formation of ozone, which can cause or aggravate respiratory illnesses.[3]

EPA began establishing NAAQS for criteria air pollutants in the early 1970s. When the NAAQS began going into effect in the 1970s, tall stacks were built in large numbers as a dispersion technique to help reduce ground-level concentrations of pollutants in the immediate vicinity of the stack. In 1970, there were only 2 stacks higher than 500 feet in the United States, but this number had increased to more than 180 by 1985.

While constructing a tall stack is a dispersion technique that helps to reduce pollution concentrations in the local area, using tall stacks does not reduce total emissions that can potentially be transported to downwind states. The 1977 amendments to the Clean Air Act discouraged the use of dispersion techniques to help attain NAAQS. Specifically, section 123 prohibits states from counting the dispersion effects of stack heights in excess of a stack's GEP height when determining a source's emissions limitation. The Clean Air Act defines GEP as "the height necessary to insure that emissions from the stack do not result in excessive concentrations of any air pollutant in the immediate vicinity of the source as a result of atmospheric downwash, eddies, or wakes which may be created by the source itself, nearby structures, or nearby terrain obstacles."[4] According to federal regulations, a stack's GEP height is the higher of

- 65 meters, measured from the ground-level elevation at the base of the stack;
- a formula based on the height and width of nearby structure(s) (height plus 1.5 times the width or height, whichever is lesser);[5] or
- the height demonstrated by a fluid model or field study that ensures the emissions from a stack do not result in excessive concentrations of any air

pollutant as a result of atmospheric downwash created by the source itself, nearby structures, or nearby terrain features.

Downwash occurs when large buildings or local terrain distort or impact wind patterns, and an area of more turbulent air forms, known as a wake.

Emissions from a stack at a power plant can be drawn into this wake and brought down to the ground near the stack more quickly (see fig. 1).

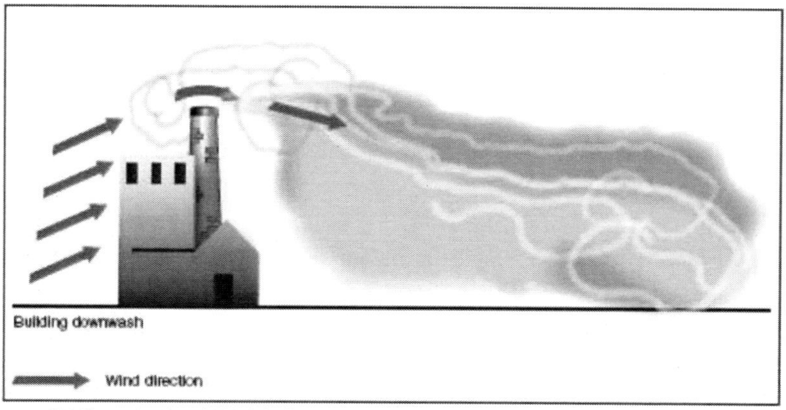

Source: GAO analysis of EPA information.

Figure 1. Building Downwash.

States issue air permits to major stationary sources of air pollution, such as power plants, and determine GEP for stacks when they set emissions limitations for these sources. Emissions limitations may be reset when plants undergo New Source Review. New Source Review is a preconstruction permitting program which requires a company that constructs a new facility or makes a major modification to an existing facility to meet new, more stringent emissions limitation based on the current state of pollution control technology. A stack's GEP height is used in air dispersion modeling that takes place when emissions limitations are developed for a source as part of the permitting process.

Many sources contribute to levels of pollution that affect the ability of downwind states to attain and maintain compliance with NAAQS, and some of these pollutants may originate hundreds or thousands of miles from the areas where violations are detected. The Clean Air Act's "good neighbor provisions" under section 110 of the Act require states to prohibit emissions that significantly contribute to nonattainment or interfere with maintenance of

NAAQS in downwind states or which will interfere with downwind states' ability to prevent significant deterioration of air quality.[6] Section 126 of the Clean Air Act also allows a downwind state to petition EPA to determine that specific sources of air pollution in upwind states interfere with the downwind state's ability to protect air quality and for EPA to impose emissions limitations directly on these sources. As detailed in the timeline below, Congress granted EPA increased authority to address interstate transport of air pollution under the Clean Air Act, and EPA acted on this authority.

- *1977 amendments to the Clean Air Act.* These amendments contained two provisions that focused on interstate transport of air pollution, the predecessor to the current good neighbor provision of section 110 of the Act and section 126. These amendments also established the New Source Review program.
- *1990 amendments to the Clean Air Act.* These amendments added the Acid Rain Program (Title IV) to the Clean Air Act, which created a capand-trade program for SO_2 emissions from power plants, with a goal of reducing annual SO_2 emissions by 10 million tons from 1980 levels and reducing annual NO_x emissions by 2 million tons from 1980 levels by the year 2000.
- *1998 NO_x SIP Call.* After concluding that NO_x emissions from 22 states and the District of Columbia contributed to the nonattainment of NAAQS for ozone in downwind states, EPA required these states to amend their SIPs to reduce their NO_x emissions. EPA took this regulatory action based on section 110 of the Clean Air Act.
- *2005 Clean Air Interstate Rule (CAIR).* This regulation required SIP revisions in 28 states and the District of Columbia that were found to contribute significantly to nonattainment of NAAQS for fine particulate matter and ozone in downwind states. CAIR required reductions for SO_2 and NO_x emissions from 28 eastern states and the District of Columbia and included an option for states to meet these reductions through regional cap-and-trade programs. When the rule was finalized, EPA estimated it would annually reduce SO_2 and NO_x emissions by 3.8 million and 1.2 million tons, respectively, by 2015. The US Court of Appeals remanded CAIR to EPA in 2008 because it found significant flaws in the approach EPA used to develop CAIR, but allowed the rule to remain in place while EPA develops a replacement rule.

- *2010 Transport Rule*. EPA proposed this rule to replace CAIR, which aims to reduce emissions of SO_2 and NO_x from power plants.[7] If finalized as written, the rule would require emissions of SO_2 to decrease 71 percent over 2005 levels and emissions of NO_x to decrease by 52 percent over 2005 levels by 2014.[8]

As described above, EPA's efforts to address the interstate transport of air pollution from power plants have focused on reducing the total emissions of SO2 and NOx from these plants. Unlike tall stacks, pollution controls help to reduce the actual emissions from power plants by either reducing the formation of these emissions or capturing them after they are formed. At coal power plants, these controls are generally installed in either the boiler, where coal is burned, or the duct work that connects a boiler to the stack. A single power plant can use multiple boilers to generate electricity, and the emissions from multiple boilers can sometimes be connected to a single stack. Figure 2 shows some of the pollution controls that may be used at coal power plants: fabric filters or electrostatic precipitators (ESP) to control particulate matter, flue gas desulfurization (FGD) units—known as scrubbers—to control SO2 emissions, and selective catalytic reduction (SCR) or selective non-catalytic reduction (SNCR) units to control NOx emissions.

The reduction in emissions from a coal power plant by the use of pollution controls can be substantial, as shown in table 1.

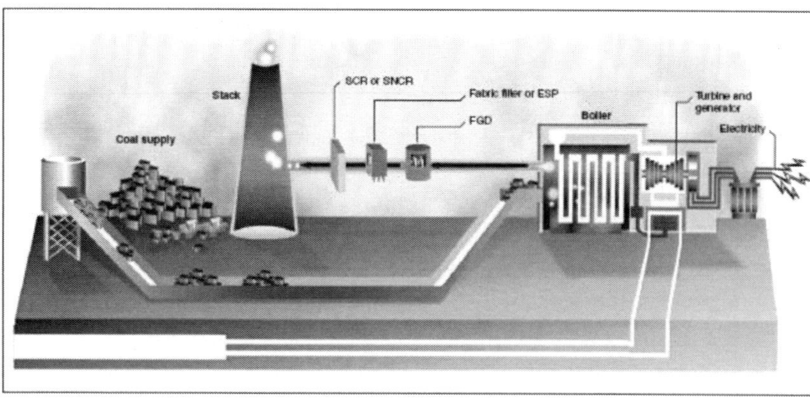

Source: GAO analysis of information from Electric Power Research Institute and Tennessee Valley Authority.

Figure 2. Sample Layout of Pollution Controls in a Coal Power Plant.

The installation of pollution control equipment can also be expensive. According to a Massachusetts Institute of Technology study of coal power plants, it may cost anywhere from \$215,000 per megawatt to \$330,000 per megawatt to install controls at a coal power plant for particulate matter, SO_2, and NO_x.[9]

For a typical coal power plant with a capacity of 500 megawatts, this means that it could cost from \$107 million to install these controls at a newly built facility up to \$165 million to retrofit these controls at an existing facility. Additionally, pollution controls can require additional energy to operate, known as an energy penalty.

Table 1. Summary of Pollution Control Equipment Used at Coal Power Plants

Pollutant targeted	Control equipment name	How it works	Removal efficiency
Particulate matter	ESP	An induced electrical charge removes particles from flue gas	Capable of 99.0-99.5% removal of particulates
Particulate matter	Fabric filter (commonly referred to as a "baghouse")	Flue gas passes through a tightly woven fabric filter	Capable of 99.9% removal of particulates
SO_2 [a]	FGD unit (commonly referred to as a "scrubber")	Wet FGDs inject a liquid sorbent, such as limestone, into the flue gas to form a wet solid that can be disposed of or sold Dry FGDs inject a dry sorbent, such as lime, into the flue gas to form a solid by-product that is collected	Wet FGDs – Capable of 80-99% removal of SO2 Dry FGDs – Capable of 70-95% removal of SO_2
NO x	Combustion control technologies, such as low-NO x burners[b]	Coal combustion conditions are adjusted so that less NOx formation occurs	Capable of 40-45% reduction in the formation of NO x
NO x	Post-combustion controls, such as SCR and SNCR units	SCRs inject ammonia into flue gas to form nitrogen and water and use a catalyst to enhance the reaction SNCRs inject ammonia as well, but do not use a catalyst	SCRs – Capable of 70-95% removal of NOx SNCRs – Capable of 30-75% removal of NOx

Source: GAO summary of reports by EPA, National Academies, Electric Power Research Institute, and industry documents.

[a] Another approach to reducing SO2 emissions from a coal power plant is for a plant to switch from using coal with a higher sulfur content to coal with a lower sulfur content, or to blend higher sulfur coal with lower sulfur coal.

[b] Low-NOx burners can be used in conjunction with post-combustion controls for NOx as well.

ALMOST 300 TALL SMOKESTACKS OPERATE IN 34 STATES, AND ABOUT HALF BEGAN OPERATING BEFORE 1980

According to our analysis of EIA data, which we updated with our survey results, we found a total of 284 tall smokestacks were operating at 172 coal power plants in 34 states, as of December 31, 2010.

While about half of the tall stacks began operating more than 30 years ago, there has been an increase in the number of tall stacks that have begun operating in the last 4 years, which several stakeholders attributed to the need for new stacks when retrofitting existing plants with pollution control equipment.

284 Tall Stacks Were Operating at about 170 Coal Power Plants, with Approximately One-Third Located in the Ohio River Valley

As of December 31, 2010, we found a total of 284 tall stacks were operating at 172 coal power plants in the United States. These tall stacks account for about 35 percent of the 808 stacks operating at coal power plants in the United States, and they are generally located at larger power plants. Specifically, we found these stacks are associated with 64 percent of the coal generating capacity.[10]

We found that 207 tall stacks (73 percent) are between 500 and 699 feet tall and that 63 stacks (22 percent) are between 700 and 999 feet tall. The remaining 14 stacks (5 percent) are 1,000 feet tall or higher, with the tallest stack at a coal power plant in the United States having a height of 1,038 feet at the Rockport Power Plant in Indiana. In figure 3, we show how a tall stack compares to the heights of other well-known structures.

Thirty-five percent of the 284 tall stacks are concentrated in 5 states along the Ohio River Valley—Kentucky, Ohio, Indiana, Illinois, and Pennsylvania—at 59 coal power plants.

Another 32 percent are located in Alabama, Missouri, West Virginia, Michigan, Georgia, Wyoming, Wisconsin, and Texas, while the remaining 33 percent of tall stacks are located across 21 other states.11 Figure 4 shows the location of coal power plants with operating tall stacks. For counts of all tall stacks by state, see appendix II.

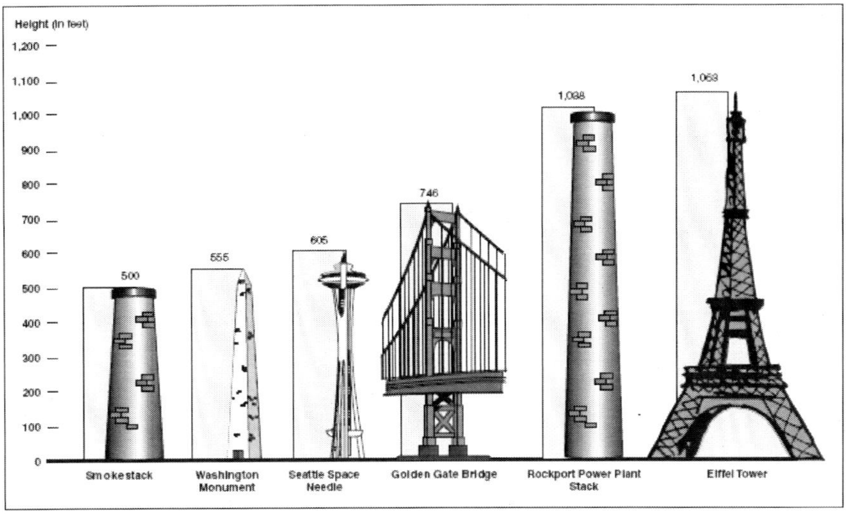

Source: GAO analysis of information for relevant buildings.

Figure 3. Comparison of Tall Stacks to Well-Known Structures.

Sources: GAO and Map Resources (map).
Note: Alaska and Hawaii are not included because they do not have tall stacks.

Figure 4. Location of Coal Power Plants with Operating Tall Stacks, as of December 2010.

About Half of All Tall Stacks Began Operating before 1980, but an Increasing Number Have Gone into Service in the Last 4 Years

Forty-six percent of the 284 tall stacks operating at coal power plants in the United States as of December 31, 2010, went into service before 1980. Another 28 percent went into service in the 1980s, 7 percent went into service in the 1990s, and 18 percent went into service since 2000. Of the stacks that went into service since 2000, a vast majority went into service in the last 4 years, as shown in figure 5.

A large majority of the tall stacks that went into service in the past 4 years are replacements of existing, older stacks. Several stakeholders told us many of these older stacks were replaced to accommodate changes in flue gas that resulted from the installation of certain types of pollution control equipment to meet emission reductions required by the first phase of CAIR. For example, stakeholders explained that a FGD unit—used to reduce SO_2 emissions—reduces the temperature and increases the moisture of a plant's flue gas.

Stack Height Contributes to Interstate Transport of Air Pollution, and the Emissions from Several Tall Stacks Remain Uncontrolled for Certain Pollutants

Stack height is one of several factors that contribute to the interstate transport of air pollution. While the use of pollution controls has increased in recent years at coal power plants, several boilers connected to tall stacks remain uncontrolled for certain pollutants.[12]

Stack Height Is One of Several Factors that Contribute to the Interstate Transport of Air Pollution

Stack height is one of several factors that contribute to the interstate transport of air pollution. According to reports and stakeholders with expertise on this topic, tall stacks generally disperse pollutants over longer distances than shorter stacks and provide pollutants with more time to react in the atmosphere to form ozone or particulate matter.

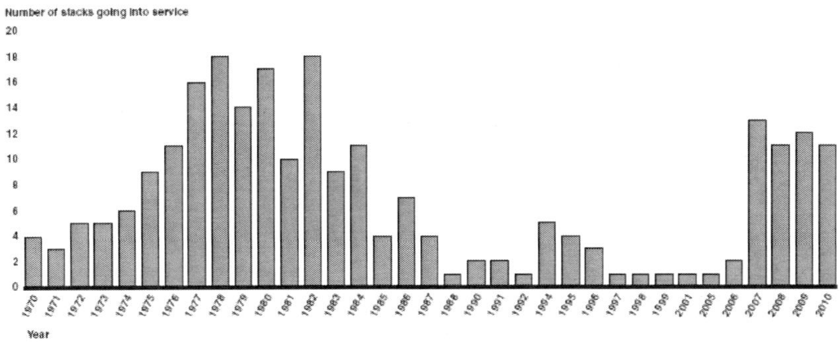

Source: GAO analysis of EIA data and survey results.

Figure 5. Distribution of Operating Tall Stacks by Year Stack Went Into Service.

However, the interstate transport of air pollution is a complex process that involves several variables—such as total emissions from a stack, the temperature and velocity of the emissions, and weather—in addition to stack height. As a result, stakeholders had difficulty isolating the exact contribution of stack height to the interstate transport of air pollution, and we found limited research on this specific topic. For example, EPA staff involved in the modeling of interstate transport told us that it is difficult to determine the different impacts that stacks of varying heights have on the transport of air pollution. According to one atmospheric scientist we spoke with, the interstate transport of air pollution is a complex process and stack height represents just one variable in this process.

Stakeholders struggled to identify the precise impact of tall stacks, due in part to the other factors that influence how high emissions from a stack will rise. The temperature and velocity of a stack's emissions, along with its height, contribute to what is known as an "effective stack height." Effective stack height takes into account not only the height at which emissions are released, but also how high the plume of emissions will rise, which is influenced by the temperature and velocity of these emissions. One atmospheric scientist told us the emissions from a shorter stack could rise higher than a taller stack, depending on the temperature and velocity of the emissions.

Weather also plays a key role in the transport of air pollution. A study by the Northeast States for Coordinated Air Use Management (NESCAUM)— a group that represents state air agencies in the Northeast—described weather patterns that can contribute to high-ozone days in the Ozone Transport Region, which includes 12 states in the Mid-Atlantic and New England regions and the District of Columbia.[13] These high-ozone days typically occur in the summer on hot days, when the sun helps transform NO_x and volatile organic compounds into ozone. Wind speeds and wind direction also help to determine how emissions will travel. In the Mid-Atlantic United States, the wind generally blows from west to east during the day, and wind speeds are generally faster at higher elevations. The time of day can also influence the transport of air pollution. According to the NESCAUM report and researchers we spoke with, ozone can travel hundreds of miles at night with the help of high-speed winds known as the low-level jet. This phenomenon typically occurs at night when an atmospheric inversion occurs due to the ground cooling quicker than the upper atmosphere. A boundary layer can form between these two air masses several hundred feet off the ground, which can allow the low-level jet to form and transport ozone and particulate matter with its high winds. As the atmosphere warms the following day, this boundary layer can break down and allow these transported emissions to mix downward and affect local air quality.

Air dispersion models typically take into account stack height along with these other factors when predicting the transport of emissions from power plants. For example, EPA used the Comprehensive Air Quality Model with Extensions (CAMx) to conduct the modeling to support the development of the Transport Rule. CAMx is a type of photochemical grid model, which separates areas into grids and aims to predict the transport of sources that lie within these grids. Key inputs into this model include stack height, the velocity and temperature of emissions, and weather data.[14] EPA staff involved in conducting this modeling for the Transport Rule said they use the CAMx model to predict the actual impacts of air emissions, and they have not used this model to estimate the specific impact of stack height on interstate transport. They reported their modeling efforts in recent years have been done in support of CAIR and the Transport Rule, and have been focused on modeling the regional impacts of reducing total air emissions.

Several stakeholders we spoke with said total emissions is a key contributor to interstate transport of air pollution, and the use of pollution controls at coal power plants is critical to reducing interstate transport of air pollution.

Reducing the total emissions from a power plant influences how much pollution can react in the atmosphere to form ozone and particulate matter that can ultimately be transported.

Use of Pollution Controls at Coal Power Plants Has Increased in Recent Years, but Emissions from Some Plants, Including Several with Tall Stacks, Remain Uncontrolled for Certain Pollutants

The use of pollution control equipment, particularly for SO_2 and NO_x emissions, has increased over time, largely in response to various changes in air regulations, according to stakeholders and reports we reviewed. According to EIA data, the generating capacity of power plants that is controlled by FGDs has increased from about 87,000 megawatts to about 140,000 megawatts from 1997 to 2008.[15] Since coal power plants had about 337,000 megawatts of generating capacity in 2008, this means that about 42 percent of the generating capacity was controlled by a FGD in 2008. Similarly, SCRs were installed at about 44,000 megawatts worth of capacity from 2004 through 2009, with about one-third of these installations occurring in 2009 alone, according to an EPA presentation on this topic. EPA and state officials, along with electric utility officials, told us that the increase in the use of these pollution controls is due to various air regulations, such as the Acid Rain Program and CAIR, which focused on reducing SO_2 and NO_x emissions.

However, while we found that the use of pollution controls at coal power plants has increased in recent years, many boilers remain uncontrolled for certain pollutants, including several connected to tall stacks. For example, we found that 56 percent of the boilers attached to tall stacks at coal power plants do not have a FGD to control SO_2 emissions. Collectively, we found that these uncontrolled boilers accounted for 42 percent of the total generating capacity of boilers attached to tall stacks.[16] Our findings on FGDs are similar to EPA data on all coal power plants. In 2009, EPA estimated that 50 percent of the generating capacity of coal power plants did not have FGDs.

For NO_x controls, we found that while about 90 percent of boilers attached to tall stacks have combustion controls in place to reduce the formation of NO_x emissions, a majority of these boilers lack post-combustion controls that can help to reduce NO_x emissions to a greater extent. Specifically, 63 percent of boilers connected to tall stacks do not have post-combustion controls for NO_x, such as SCRs or SNCRs, which help reduce NO_x emissions more than combustion controls alone. Collectively, we found that these boilers without post-combustion controls accounted for 54 percent of the total generating capacity of boilers attached to tall stacks. EPA data on all coal power plants show that 53 percent of the generating capacity for coal power plants did not have post-combustion controls for NO_x emissions in place in 2009.

Tall stacks that had uncontrolled SO_2 and NO_x emissions were generally attached to older boilers that went into service prior to 1980. We found that approximately 85 percent of boilers without FGDs that were attached to tall stacks went into service before 1980. Similarly, over 70 percent of the boilers without post-combustion controls for NO_x went into service before 1980. Overall, we found that about 82 percent of the boilers that lacked both a FGD and post-combustion controls for NO_x went into service before 1980. Some stakeholders attributed the lack of pollution controls on older boilers to less stringent standards that were applied at the time the boilers were constructed. As discussed above, companies that construct a new facility or make a major modification to an existing facility must meet new emissions limitations based on the current state of pollution control technology. Because pollution control technology has advanced over time, the standards have become more stringent over time, meaning that boilers constructed before 1980 would have had higher allowable emissions and less need to install controls than boilers constructed in 2010.

Unlike our findings on FGDs and post-combustion controls for NO_x emissions, we found that 100 percent of boilers attached to tall stacks were controlled for particulate matter. However, it is important to note that plants with uncontrolled SO_2 and NO_x emissions contribute to the formation of additional particulate matter in the atmosphere.

Based on Available Information, 17 of 48 Tall Smokestacks Built Since 1988 Exceed Their GEP Height, and a Variety of Factors Can Influence Height Decisions

We identified 48 tall stacks built since 1988 that states reported are subject to the GEP provisions of the Clean Air Act and for which states could provide GEP height information. Of these 48 stacks, we found that 17 exceed their GEP height, 19 are at their GEP height, and 12 are below their GEP height. We found that 15 of the 17 stacks built above GEP were replacement stacks that were built as part of the process of installing pollution control equipment. These stacks vary in the degree to which they exceed GEP height, ranging from less than 1 percent above GEP to about 46 percent above GEP, as shown in table 2. The other 2 stacks built above GEP exceed their GEP height by 2 percent or less.

Table 2. Stacks Built Since 1988 with Heights that Exceed GEP

Plant name and unit number	State	In-service date (year)	Stack height (feet)	GEP height (feet)	Percentage difference between actual and GEP height	Replacement stack?
Bowen (units 3, 4)	Ga.	2008	675	643	5	Yes
Bowen (units 1, 2)	Ga.	2009	675	643	5	Yes
Hammond (units 1, 2, 3, 4)	Ga.	2008	675	464	46	Yes
Wansley (units 1, 2)	Ga.	2008	675	663	2	Yes
Duck Creek (unit 1)	Ill.	2008	588	533	10	Yes
Paradise (unit 3)	Ky.	2006	600	420	43	Yes
Iatan (unit 2)	Mo.	2010	605	604	0.2	Yes
Miami Fort (unit 7)	Ohio	2007	800	705	14	Yes
Miami Fort (unit 8)	Ohio	2007	800	590	36	Yes
WH Sammis (units 1, 2, 3, 4, 5, 6, 7)	Ohio	2010	850	840	1	Yes
Homer City (unit 3)	Pa.	2001	864	853	1	Yes
Montour (units 1, 2)	Pa.	2008	700	540	30	Yes
Brunner Island (unit 3)	Pa.	2009	600	540	11	Yes
Bull Run (unit 1)	Tenn.	2008	500	492	2	Yes
Fayette (unit 3)	Tex.	1988	535	533	0.4	No
JK Spruce (unit 2)	Tex.	2009	601	588	2	No
Mountaineer (unit 1)	W. Va.	2007	1,000	839	19	Yes

Source: GAO analysis of state survey responses.

When we followed up with utility officials regarding why these stacks were built above GEP, they reported that a variety of factors can influence stack height decisions. These factors included helping a plant's emissions clear local geographic features, such as valley walls.[17] According to one state air protection agency, three stacks were built above GEP to provide further protection against downwash. Officials from two utilities said they built stacks above GEP at coal power plants to account for the impact of other structures, such as cooling towers, on the site.[18] Other stakeholders said that utilities may be hesitant to lower stack heights at their facilities when replacing a stack because plant officials have experience with that stack height and its ability to help protect against downwash. An official from one company that builds stacks told us this practice has sometimes occurred because utilities do not want the moisture-rich emissions from the replacement stack to hasten the deterioration of the old stacks, which are usually left in place and must be maintained. In addition, this moisture can create large icicles on the older stacks, which can present a danger to staff working at the power plant.

Other stakeholders highlighted factors that may play a role in making stack height decisions. Some federal and state officials reported that generally there is little incentive to build a stack above GEP because a facility will not receive dispersion credit for the stack's height above GEP. Other stakeholders acknowledged that a stack could be built above GEP for site-specific reasons, such as helping emissions clear nearby terrain features. Some of these officials also noted that cost was another factor considered when making stack height decisions, as it is generally more costly to build a higher stack. For example, one utility official told us that two replacement stacks that were recently built below their original heights could meet their emissions limitations with these lower stack heights because the utility was installing pollution control equipment and did not want to incur the additional cost of building a taller stack.

We found that stacks built above GEP since 1988 generally were attached to boilers that had controls in place for SO_2, NO_x, and particulate matter, as shown in table 3. We found similar results for stacks that were built at or below their GEP heights.

Table 3. Information on Pollution Controls for Boilers Attached to Tall Stacks Built Since 1988 for which GEP Information was Available

Stacks	Proportion of attached boilers with a FGD installed for SO_2 emissions	Proportion of attached boilers with combustion controls installed for NO emissions	Proportion of attached boilers with post-combustion controls installed for NO emissions	Proportion of attached boilers with controls installed for particulate matter
Stacks built above GEP	100%	97%	80%	100%
Stacks built at GEP	82	93	75	100

Source: GAO analysis of EIA data and survey results.

We were unable to obtain GEP height information for an additional 25 stacks that were built since 1988 for two reasons. First, some of these stacks replaced stacks that were exempt from the GEP regulations, according to state officials. Section 123 of the Clean Air Act exempts stack heights that were in existence on or before December 31, 1970, from the GEP regulations; because the exemption applies to stack heights rather than to stacks themselves, it covers both original and replacement stacks.[19] Second, states did not have GEP information readily available for some stacks. According to state officials, they did not set new emissions limits at the time these replacement stacks were built because they were part of pollution control projects and emissions from these plants did not increase. For example, one state reported that GEP could have been calculated decades earlier for the original stacks when emissions limitations were set, and they were unable to locate this information in response to our request.20 According to EPA staff we spoke with about this issue, states are not required to conduct a GEP analysis in these instances. While we were unable to obtain GEP information for these stacks, our analysis of the pollution controls installed at boilers connected to these stacks yielded similar results to those stacks for which we did obtain GEP information. Specifically, all of these boilers had controls for SO2 and particulate matter, and 85 percent had post-combustion controls for NOx.

AGENCY COMMENTS

We provided a draft of this report to EPA and DOE for review and comment. Both EPA and DOE stated they had no comments.

Sincerely yours,
David C. Trimble Acting Director
Natural Resources and Environment

APPENDIX I: SCOPE AND METHODOLOGY

To identify the number and location of smokestacks at coal power plants that were 500 feet or higher as of December 31, 2010, we analyzed data on power plants from the Department of Energy's (DOE) Energy Information Administration (EIA). We also used these data to determine when these stacks began operating. To determine the reliability of these data, we reviewed documentation from EIA, interviewed relevant officials who were involved in collecting and compiling the data, conducted electronic testing of the data, and we determined that the data were sufficiently reliable for our purposes. Because the EIA data were collected in 2008, we contacted all 50 states and the District of Columbia to determine if they had tall stacks and developed and administered a survey to those 38 states with tall stacks to update the relevant EIA data and determine if any changes had taken place in the number or operating status of stacks since that time.[1] We received e-mail addresses for each state from the Web site of the National Association of Clean Air Agencies, which represents air pollution control agencies in 53 states and territories, and developed a survey that we sent to respondents as an e-mail attachment. Prior to sending out this survey, we pretested the survey with officials from 2 states and revised some of the survey questions based on their input. We received responses to our survey from all 38 states and we sent follow-up questions based on their survey responses to clarify certain responses or to ask for additional information. We updated the relevant EIA data with these survey results to include the most recent information available on tall stacks.

We did not include tall stacks that were used as bypass stacks only in times of maintenance or emergencies in our count of tall stacks. State officials reported that bypass stacks are rarely used and would not be used at the same time as plants' fully operating stacks. Additionally, we defined multi-flue stacks—those with multiple flues running within a single casing—as one stack, as opposed to counting each flue as a separate stack. A state modeling official told us they consider multi-flue stacks as single stacks when conducting dispersion modeling.

For the purposes of this report, we defined tall smokestacks to be those that were 500 feet or higher. In our interviews with stakeholders, several told us they considered 500 feet to be a "tall" stack. Some stakeholders said that a typical boiler building at a coal power plant is about 200 feet high. Given that the original formula for good engineering practice (GEP) was 2.5 times height of nearby structures, this would equal about 500 feet. Other stakeholders reported that they considered a stack built above GEP to be "tall."

To determine what is known about tall stacks' contribution to the interstate transport of air pollution, we reviewed reports from the Environmental Protection Agency (EPA) and academics and spoke with stakeholders with expertise on this topic. We conducted a literature search of engineering and other relevant journals on the topic of stack height and interstate transport of air pollution, and we reviewed the limited amount of literature we identified. The stakeholders we interviewed included EPA officials involved in modeling interstate transport of air pollution from power plants, officials from utilities and construction firms that design and build power plants, atmospheric scientists who conduct research on this topic, and state officials who are involved in permitting power plants and complying with federal regulations governing the interstate transport of air pollution. We also analyzed the EIA data and our survey results to determine the pollution control equipment installed at coal power plants with stacks 500 feet or higher. Specifically, we identified the control equipment that was associated with boilers that were attached to tall stacks. Pollution control equipment is not installed on stacks themselves; rather it is installed in the boilers or the ductwork that connect the boiler to a stack. We also interviewed stakeholders to learn about trends in installing pollution control equipment and reviewed relevant reports on this topic.

To determine the number of tall stacks that have been built above their GEP height since 1988, we used our survey to obtain information from state officials about the GEP height for these stacks. Twenty-two states had stacks that were over 500 feet that were built since 1988, and we received survey responses from all of them. In our survey, we also asked for reasons that a stack was built above GEP, when applicable. In cases where state officials could not provide specific reasons, we contacted the utilities that operate the plants with these stacks to obtain this information. Specifically, we contacted utilities that were involved in operating 15 of the 17 stacks that were built since 1988 and exceed GEP height, and we were able to interview utilities operating 12 of these stacks. We did not contact the utilities that operate the other 2 stacks, because the stacks are each less than 2 feet above GEP. We also interviewed companies that design and build power plants to ask about some of the general factors that are considered when deciding on stack height. We focused on stacks built since 1988, because that was the year that EPA's regulations for determining GEP height were largely affirmed by the District of Columbia Court of Appeals. EPA began the process of developing these regulations in the late 1970s, but the final regulations were not issued until 1985. The regulations were then challenged in court and were largely affirmed in 1988.

Finally, we conducted site visits to two coal power plants in Ohio. We selected this state because it contained several coal power plants with tall stacks, including some stacks that were built in 1988 or later. During this visit, we interviewed utility officials that operated these plants, along with state and local officials involved in permitting these plants.

We conducted this work from July 2010 through May 2011 in accordance with all sections of GAO's quality assurance framework that are relevant to our objectives. This framework requires that we plan and perform the engagement to obtain sufficient, appropriate evidence to meet our stated objectives and to discuss any limitations in our work. We believe that the information and data obtained, and the analysis conducted, provide a reasonable basis for any findings and conclusions.

APPENDIX II: DISTRIBUTION OF TALL STACKS BY STATE

Table 4 provides counts of the number of stacks 500 feet or higher—tall stacks—by state. In addition, the table provides information on the generating capacity of the boilers attached to these stacks.

Table 4. Number of Tall Stacks at Coal Power Plants by State and Associated Generating Capacity of Boilers Attached to These Stacks

State	Number of tall stacks	Generating capacity (megawatts)
Ohio	22	19,626
Kentucky	22	14,491
Indiana	19	14,286
Illinois	19	11,824
Pennsylvania	17	15,765
Alabama	14	11,664
Missouri	12	9,360
West Virginia	12	13,920
Michigan	12	8,971
Georgia	11	13,793
Texas	11	9,277
Wyoming	10	4,486
Wisconsin	10	5,264
Arizona	9	4,704
Colorado	8	3,820
Utah	7	4,608
Oklahoma	7	4,112
Florida	7	5,720
Minnesota	6	4,395
Tennessee	6	6,292
North Dakota	6	2,997
Louisiana	5	3,207
Kansas	5	3,738
Iowa	4	3,187
Montana	4	2,272
North Carolina	4	3,404
Arkansas	3	3,958
Nebraska	3	2,014
South Carolina	3	1,564
Nevada	2	572
Delaware	1	164
Oregon	1	601

State	Number of tall stacks	Generating capacity (megawatts)
Maryland	1	728
New York	1	655
Total	284	215,439

Source: GAO analysis of EIA data and survey results.

End Notes

[1] For the purposes of this report, we consider tall smokestacks to be those that are 500 feet or higher.

[2] Federal regulations further define GEP as the higher of 65 meters (about 213 feet), the results of a calculation based on the dimensions of nearby structure(s), or the results of a fluid modeling demonstration. The calculation based on the dimensions of nearby structure(s) that applies to stacks built after January 12, 1979, states that GEP = H + 1.5 L, where H is equal to the height of nearby structure(s) and L is equal to the height or width of nearby structure(s), whichever is less. For stacks built since December 31, 1970, and in existence on January 12, 1979, this calculation is GEP = 2.5H, where H is equal to the height of nearby structure(s).

[3] Ozone is formed through a series of chemical reactions between NOx; other chemicals in the atmosphere, known as volatile organic compounds; and sunlight. Cars and power plants that burn fossil fuels are contributors of NOx pollution.

[4] 40 USC. § 7423(c) (2006). GEP is a regulatory term used to refer to the minimal height necessary to avoid excessive downwash, but does not necessarily imply that the GEP height is optimized based on structural engineering principles.

[5] For stacks in existence on January 12, 1979, and for which the owner or operator had obtained all applicable permits or approvals, the GEP height formula is 2.5 times the height of nearby structure(s). Structures that are next to one another are considered a single structure if their "distance of separation is less than their smallest dimension (height or width)." See EPA, Guidelines for Determination of Good Engineering Practice Stack Height (Research Triangle Park, N.C., 1985).

[6] Prevention of significant deterioration is a standard used to refer to areas of the country which are already in attainment with NAAQS. Sources that are constructed or undergo major modifications in such areas must install the Best Available Control Technology to help prevent the air quality from deteriorating to the level set by NAAQS.

[7] EPA believes that the Transport Rule addresses the court's concerns with CAIR by, among other things, introducing a state-specific methodology for identifying significant contributions to nonattainment and interference with maintenance, and proposing remedy options to ensure that all necessary reductions are achieved in the covered states.

[8] In particular, the Transport Rule focuses on helping states attain the 8-hour ozone standard and the particulate matter 2.5 standard. This particulate matter 2.5 standard focuses on particles that are 2.5 micrometers in diameter and smaller, about 1/30th the diameter of a human hair, which have been shown to aggravate respiratory and cardiovascular disease.

[9] Massachusetts Institute of Technology, The Future of Coal (Cambridge, Mass., 2007).

[10] For the purposes of this report, our discussion of capacity refers to nameplate capacity, which refers to maximum rated output of electric generating units as designed by the manufacturer.

[11] Percentages may not sum to 100 due to rounding.

[12] Pollution control equipment is not installed directly on smokestacks. Instead, pollution control equipment is installed throughout a power plant to reduce the formation of pollutants and remove them before they are emitted through the stack.

[13] NESCAUM, The Nature of the Ozone Air Quality Problem in the Ozone Transport Region: A Conceptual Description (Boston, Mass., August 2010). The Ozone Transport Region of the eastern United States covers over 62 million people living in Connecticut, Delaware, the District of Columbia, Maine, Maryland, Massachusetts, New Hampshire, New Jersey, New York, Pennsylvania, Rhode Island, Vermont, and northern Virginia.

[14] There are also other air models that are used to predict the dispersion of air pollution from specific sources of emissions, such as AERMOD. The data inputs into AERMOD include 5 years of regionally representative meteorological data, emissions rate, and stack parameters (height, velocity, and temperature of emissions).

[15] A megawatt is a unit for measuring the electric generation capacity of a power plant. One megawatt of capacity operating for 1 full day produces 24 megawatt-hours—or 24,000 kilowatt-hours of electricity. According to EIA analysis, the typical American home consumes about 11,040 kilowatt-hours of electricity a year.

[16] For the purposes of this report, we refer to the capacity of boilers based on information from the EIA-860 form. This form provides capacity for generators that are associated with boilers at power plants.

[17] GEP regulations permit nearby terrain features to be taken into account, but only when determining GEP through the fluid model approach. These terrain features must generally be within 0.5 mile to 2 miles of the source being modeled. If GEP is calculated using only the dimensions of nearby buildings, such nearby terrain features are not part of this calculation. See EPA, Guidelines for Determination of Good Engineering Practice Stack Height.

[18] The guidance on determining GEP permits aerodynamic structures such as hyperbolic cooling towers to be taken into account, but only when determining GEP through the fluid model approach. See EPA, Guidelines for Determination of Good Engineering Practice Stack Height.

[19] The continued exemption of a stack also depends on whether the source using the stack is reconstructed or undergoes a major modification. Section 123 also exempts federal coalfired power plants which commenced operations before July 1, 1957, and whose stacks were constructed under a contract awarded before February 8, 1974.

[20] As we noted earlier, the Clean Air Act began requiring the incorporation of emissions limitations into operating permits in 1990. Prior to the 1990 amendments to the Clean Air Act, information applicable to individual sources, including the source's pollution control obligations, was not collected in a single permit but could be scattered throughout numerous provisions of the SIP. This fact may have contributed to officials' difficulty in finding information on sources for which GEP was calculated at an earlier time.

End Note for Appendix I

[1] Our survey of 38 states included some stacks that were at power plants that were not fueled by coal or that were not operating yet. When we restricted our analysis to operating tall stacks at coal power plants, we found that such stacks were operating in 34 states.

In: Federal Control of Air Emissions
Editors: H.M. Graham and T.E. Lee
ISBN: 978-1-62081-808-4

Chapter 4

THE NATIONAL AMBIENT AIR QUALITY STANDARDS (NAAQS) FOR PARTICULATE MATTER (PM): EPA'S 2006 REVISIONS AND ASSOCIATED ISSUES*

Robert Esworthy and James E. McCarthy

SUMMARY

Following its review of more than 2,000 scientific studies, on October 17, 2006, the Environmental Protection Agency (EPA) published its final revisions to the National Ambient Air Quality Standards (NAAQS) for particulate matter (particulates, or PM). Several states and industry, agriculture, business, and environmental and public health advocacy groups petitioned the US Court of Appeals for the District of Columbia Circuit, challenging certain aspects of EPA's revisions. A February 24, 2009, decision by the D.C. Circuit granted the petitions in part, denying other challenges, and remanded the standards to EPA for further consideration. While the court did not specifically vacate the 2006 PM standards, EPA initiated its next round of the periodic review of the particulates NAAQS, in part, in response to the court's decision, and plans to propose its decision regarding changes to the PM standards sometime during the fall of 2011. These actions, and the ongoing

* This is an edited, reformatted and augmented version of a Congressional Research Service publication, CRS Report for Congress RL34762, dated October 17, 2011.

implementation of the 2006 PM NAAQS, have prompted renewed interest among Members of Congress.

Experiences and issues leading up to and following the promulgation of the 2006 PM2.5 NAAQS could provide relevant insights as EPA proceeds with its current review. Although a tightening of the standards, the 2006 particulates NAAQS are not as stringent as recommended by EPA staff or the independent scientific advisory committee mandated under the Clean Air Act (Clean Air Scientific Advisory Committee, or CASAC). The divergence from the CASAC's recommendations proved controversial, as did several other elements of the 2006 particulates NAAQS, including the decision not to exclude rural sources from the coarse particle standard.

EPA found that the evidence continued to support associations between exposure to particulates in ambient air and numerous health problems. Based on several analytical approaches, EPA estimated that compliance with the revised NAAQS would prevent 1,200 to 13,000 premature deaths annually, as well as substantial numbers of hospital admissions and missed work days due to illness. EPA revised the PM NAAQS by strengthening the 1997 standard for "fine" particulate matter 2.5 micrometers or less in diameter (PM2.5). Specifically, the agency lowered the allowable daily concentration averaged over 24-hour periods of PM2.5 in the air from 65 micrograms per cubic meter (μg/m3) to 35 (μg/m3). The annual PM2.5 standard, which is set in addition to the daily standard to address human health effects from chronic exposures to the pollutants, was unchanged from the 1997 standard. The decision not to tighten the annual standard was overturned by the D.C. Circuit and remanded to EPA for consideration.

The 2006 particulates NAAQS also retained the 24-hour standard and revoked the annual standard for slightly larger, but still inhalable, particles less than or equal to 10 micrometers (PM10). EPA abandoned its proposal to replace the particle size indicator of PM10 with a range of 10 to 2.5 micrometers (PM10-2.5). The D.C. Circuit's February 24, 2009, decision upheld EPA's decisions with regard to PM10 NAAQS.

EPA's ongoing implementation of the 2006 NAAQS, including EPA's November 13, 2009, final designation of those geographical areas not in compliance (typically defined by counties or portions of counties), has been an area of debate among some Members of Congress, states, and other stakeholders. Although EPA did not require new nonattainment designations for PM10, the tightening of the PM2.5 standard resulted in an increased number of areas in nonattainment compared to the designations for the 1997 PM NAAQS. EPA's November 2009 final designations for the 2006 PM NAAQS included 120 counties and portions of counties in 18 states as nonattainment areas based on 2006 through 2008 air quality monitoring data.

INTRODUCTION

EPA has identified and promulgated National Ambient Air Quality Standards (NAAQS) under the Clean Air Act (CAA)[1] for six principal pollutants commonly referred to as "criteria pollutants": particulate matter (PM), ozone (O3, a key measure of smog), nitrogen dioxide (NO2, or, inclusively, nitrogen oxides,[2] NOx), sulfur oxides (SOx, or, specifically, SO2), carbon monoxide (CO), and lead (Pb). On October 17, 2006, EPA published its revisions to the NAAQS for particulates to provide protection against potential health effects associated with short- and longterm exposure to particulate matter (including chronic respiratory disease and premature mortality).[3]

EPA's revised particulates NAAQS modified the standards established in 1987[4] that focused on particles smaller than 10 micrometers (PM10, or coarse particles), and standards for "fine" particles smaller than 2.5 micrometers[5] ($PM_{2.5}$) introduced for the first time with the promulgation of the 1997 $PM_{2.5}$ NAAQS.[6] The 2006 revisions to the particulates NAAQS were the culmination of EPA's statutorily required periodic review,[7] based on its evaluation and analysis of more than 2,000 scientific studies available between 1997 and 2002, and on determinations made by the Administrator. EPA's review of the particulates NAAQS and of the scientific criteria for setting the standards completed in 2006 was initiated not long after the 1997 promulgation.

The 2006 particulates NAAQS generated national interest and debate, and oversight in Congress, as did the previous changes to the particulates standards promulgated in 1997. While the 2006 particulates NAAQS generally tightened the air quality standards for fine particulate matter, the action caused considerable controversy, including concerns that the standards were outside the range recommended by both EPA staff and by the scientific advisory panel (Clean Air Scientific Advisory Committee, or CASAC[8]) established by the CAA.[9] Conversely, some contended that available data did not support the need for stricter standards or, in some cases, the standards as promulgated in 1997.

A February 24, 2009, decision by the US Court of Appeals for the D.C. Circuit[10] to grant, in part, petitions challenging certain aspects of EPA's revised NAAQS generated renewed interest in the PM NAAQS. In December 2006, 13 states, the District of Columbia, and several state agencies had petitioned the US Court of Appeals for the D.C. Circuit to review the 2006 particulates NAAQS.[11] In addition, several groups representing various industry and agriculture interests (including coal, iron, steel, and corn refiners, oilseed processors, farmers, and cattle and pork producers), as well as environmental and public health organizations, filed petitions with the D.C. Circuit challenging the 2006 NAAQS.[12] Petitioners, joined by several interveners and amici, challenged both the fine and coarse particulates standards.

The court's decision did not vacate the PM standards but remanded certain aspects of the annual PM2.5 standard to EPA for reconsideration; the petitions did not challenge the tightening of the daily (24-hour) standard. Although implementation of the 2006 PM NAAQS continues,[13] EPA's eventual actions in response to the court's decision could have implications in the near future.

EPA's next round of the periodic review of the particulates NAAQS (including both $PM_{2.5}$ and PM10) is under way. The agency announced its intention to accelerate the review, in part in response to the D.C. Circuit Court's decision. Most recently, the agency has indicated it hoped to propose its decision regarding changes to the PM standards by the fall of 2011. Final revised standards would not likely be promulgated before the fall of 2012, based on previous EPA target dates and timelines.[14]

Released at the end of April 2011, a final EPA staff policy assessment concluded that the evidence reviewed and analyzed generally supported consideration of tightening the PM2.5 standards. The assessment also concluded that it would be appropriate to consider retaining or revising the PM10 standard depending on the weight placed on the evidence and associated uncertainties.[15] The assessment summarizes EPA staff review of the adequacy of the current standards and their analyses of the scientific evidence, quantitative assessments and air quality analyses in support of potential alternative policy options for consideration. Findings and policy options presented for consideration in EPA's assessments,[16] as well as recommendations by the Clean Air Scientific Advisory Committee (CASAC) based on their reviews of the agency staff draft assessments,[17] have generated considerable comment and debate. The issue of whether the particulates NAAQS should be strengthened or not has prompted increased scrutiny of EPA and states' implementation of the 2006 $PM_{2.5}$ NAAQS.

The range of issues raised prior to and shortly after the promulgation of the 2006 $PM_{2.5}$ NAAQS could provide relevant insights as EPA proceeds with its current review of the particulates NAAQS. This CRS report provides an analysis of the agency's final 2006 revisions to the particulates NAAQS, and an overview of the estimated costs and benefits of the revised standards and of more stringent alternatives analyzed. The report highlights concerns and issues raised regarding the 2006 revisions to the particulates standards, including those of the science advisory committee (CASAC), and actions in Congress.

EPA'S 2006 CHANGES TO THE PARTICULATES NAAQS

Establishing NAAQS does not directly limit emissions; rather, it represents the EPA Administrator's formal judgment regarding the level of ambient pollution that will protect public health with an *adequate margin of safety*. Under Sections 108-109 of the CAA, Congress mandated that EPA set national ambient (outdoor) air quality standards for pollutants whose emissions "may reasonably be anticipated to endanger public health (primary standards) or welfare[18] (secondary standards)" and "the presence of which in the ambient air results from numerous or diverse mobile or stationary sources." The statute also requires that every five years EPA review the latest scientific studies and reaffirm or modify previously established NAAQS.

The 2006 Particulates NAAQS

EPA's review of PM found that the scientific evidence since 1997 reinforced the associations between exposure to particulates and numerous cardiovascular and respiratory health problems, including aggravated asthma, chronic bronchitis, reduced lung function, irregular heartbeat, nonfatal heart attacks, and premature death. The CASAC found that the numerous epidemiological studies EPA reviewed "have shown statistically significant associations between the concentrations of ambient air $PM_{2.5}$ and PM_{10} (including levels that are lower than the 1997 particulates NAAQS) and excess mortality and morbidity."[19] Further, EPA concluded, and most of the CASAC panel concurred, that the scientific evidence supported modifying the particulates standards.

The 1997 primary NAAQS for both $_{PM2.5}$ ^{and} $_{PM10}$ included an *annual* and a *daily* (24-hour) limit. To attain the annual standard, the three-year average of the weighted annual arithmetic mean PM concentration at each monitor within an area must not exceed the maximum limit set by the agency. The 24-hour standards were a concentration-based percentile form, indicating the percentage of the time that a monitoring station can exceed the standard. For example, a 98th percentile 24-hour standard indicates that a monitoring station can exceed the standard 2% of the days during the year.

As modified and published in the October 17, 2006, *Federal Register* Notice, the primary PM2.5 and PM_{10} standards are as follows:

1. *$PM_{2.5}$*: strengthened the *daily* (24-hour) standard, which had allowed no more than 65 micrograms per cubic meter ($\mu g/m^3$), by setting a new limit of 35 $\mu g/m^3$, based on the three-year average of the 98th percentile of 24-hour PM2.5 concentrations; retained the *annual* standard at 15 $\mu g/m^3$.
2. *PM_{10}*: retained the *daily* standard at 150 $\mu g/m^3$ but changes from the 99th percentile to no more than one exceedance per year on average over three years; eliminated the *annual* maximum concentration (50 $\mu g/m^3$) standard for PM10.[20]

For $PM_{2.5}$ and $PM_{10,}$ the secondary (welfare) NAAQS are the same as the primary standards. *Table 1* below provides a comparison of the revised primary NAAQS with those previously promulgated for both $PM_{2.5}$ and PM_{10}.

Table 1. Primary (Health) NAAQS for $PM_{2.5}$ and PM_{10}: Final Revisions (2006), and Previously Promulgated

	Previous NAAQS	EPA Final Rule (2006)
	$PM_{2.5}$ (Fine)	
24-Hour Primary Standard	65 µg/m3	35 µg/m3
Annual Primary Standard	15 µg/m3	15 µg/m3
	PM_{10} (Coarse)	
24-Hour Primary Standard	150 µg/m3	150 µg/m3
Annual Primary Standard	50 µg/m3	Revoked

Source: Prepared by the Congressional Research Service with information from EPA's final particulates NAAQS (71 Federal Register 61143-61233, October 17, 2006), and related technical documents, available at http://www.epa.gov/air/particles/ actions.html. EPA's final PM staff paper and the CASAC review of the EPA staff paper are available at http://www.epa.gov/ttn/naaqs/standards/ pm/s_pm_ index. html.

EPA's final revisions to the standards for fine particulates ($PM_{2.5}$) were the same as those the agency had proposed in January 2006. However, the 2006 EPA revisions to the $PM_{2.5}$ NAAQS, while tightening the standards, were not as stringent as those recommended by the CASAC and by EPA staff.

With regard to coarse particulates, EPA had proposed replacing the current particle size indicator of PM_{10} with a range of 10 to 2.5 micrometers ($PM_{10\text{-}2.5}$), referred to as inhalable (or thoracic) coarse particles, and setting a $PM_{10\text{-}2.5}$ *daily* standard of 70 $\mu g/m^3$ rather than the current PM10 daily standard of 150 $\mu g/m^3$. The proposal also included narrowing the focus of the PM10-2.5 standard to "urban and industrial" sources and excluding particles typical to rural areas, including "windblown dust and soils and particulates generated by agricultural and mining sources." The range of alternative standards considered and proposed and issues associated with EPA's final decisions are discussed later in this report.

Implementing the 2006 PM NAAQS[21]

Promulgation of NAAQS sets in motion a process under which the states and EPA first identify geographic nonattainment areas, those areas failing to comply with the NAAQS based on monitoring and analysis of relevant air quality data.[22]

Within three years of issuance of a NAAQS, states are required to submit "infrastructure" plans demonstrating that they have the basic air quality management components necessary to implement the NAAQS.[23] Following EPA's final designations of attainment and nonattainment areas, states must submit their plans (State Implementation Plans, or SIPs) for how they will achieve and/or maintain attainment of the standards to EPA. In addition to requiring states to submit implementation plans, EPA acts to control NAAQS pollutants through national standards. These may be in the form of regulations of products that might emit the pollutants (particularly fuels and mobile sources, such as automobiles and trucks) and in the form of emission standards for new stationary sources (e.g., utilities, refineries).

The 2006 tightening of the $PM_{2.5}$ standards resulted in an increase in the number of areas (typically defined by counties or portions of counties) designated nonattainment.

EPA is not requiring new nonattainment designations for PM_{10} since the standards were not strengthened (see map in *Figure 1*, later in this report, depicting PM_{10} nonattainment areas).

2006 $PM_{2.5}$ 24-Hour NAAQS Area Designations

On November 13, 2009, EPA published its designations of 31 areas in 18 states, comprising 120 counties (89 counties and portions of 31 additional counties) for nonattainment of the revised 2006 *24-hour* $PM_{2.5}$ standard.[24] The designations, based on 2006 through 2008 air quality monitoring data, include a few counties that would be designated nonattainment for $PM_{2.5}$ for the first time, but the majority of the counties identified overlap with EPA's final nonattainment designations for the 1997 $PM_{2.5}$ NAAQS.[25] However, most of the 1997 PM2.5 nonattainment areas were *only* exceeding the annual standard; thus, tightening the 24-hour standard resulted in an increased number of areas being designated nonattainment based on exceedances of both the 24- hour *and* the annual standard. The final nonattainment designation rule became effective December 14, 2009 (30 days from the date of publication).[26] Publication of a final area designation rule for the 2006 24-hour $PM_{2.5}$ NAAQS had been delayed as a result of the incoming Administration's review of the final rule, along with several other agency proposed and final actions introduced toward the end of the previous Administration.[27]

State Implementation Plans (SIPs)

Under the CAA, within three years of issuance of a NAAQS, all states are required to submit "infrastructure" plans demonstrating that they have the basic air quality management components necessary to implement the NAAQS.[28] Areas designated attainment/unclassifiable will not have to take steps to improve air quality, but under the statute they must take steps to prevent air quality from deteriorating to unhealthy levels. For those areas designated nonattainment, state, local, and tribal governments must outline detailed control requirements in plans demonstrating how they will meet the 2006 24-hour $PM_{2.5}$ NAAQS. These plans, defined as state implementation plans and referred to as SIPs (TIPs for tribal implementation plans), must be submitted to EPA three years after the December 14, 2009, effective date of the agency's final designations.[29] If states fail to develop an adequate implementation plan, EPA can impose one. Under CAA, states are required to meet the 2006 PM2.5 standard "as expeditiously as practicable," but no later than five years from the effective date of designation—December 14, 2014, according to EPA's timeline—unless an extension (up to five additional years) allowed under the CAA is granted.[30]

For a discussion of the status of states' infrastructure, and nonattainment area SIPs for the 2006 $PM_{2.5}$ NAAQS, see CRS Report R40096, *2006 National*

Ambient Air Quality Standards (NAAQS) for Fine Particulate Matter (PM2.5): Designating Nonattainment Areas, by Robert Esworthy.

National Regulations

EPA expects that in many cases implementing national strategies—including the 1999 visibility protection regulations (Regional Haze Rule);[31] voluntary diesel engine retrofit programs; and federal emission standards for cars, light trucks, heavy-duty, and nonroad diesel engines—would provide a framework for achieving attainment with the $PM_{2.5}$ NAAQS. A key federal regulation designed to control emissions of air pollution that causes air quality problems in downwind states, EPA's May 2005 Clean Air Interstate Rule (CAIR),[32] was vacated in a July 11, 2008, decision (*North Carolina v. EPA*), by the US Court of Appeals for the D.C. Circuit.[33]

On August 8, 2011, EPA published a final rule replacing CAIR, the Cross-State Air Pollution Rule (hereinafter the "Cross-State Rule" or CSAPR).[34] The Cross-State Rule leaves the CAIR Phase 1 limits in place and sets new limits replacing CAIR's second phase in 2012 and 2014, up to three years earlier than CAIR would have.

The CSAPR covers 28 eastern, midwestern, and southern states (adding Oklahoma, Kansas, and Nebraska to the 28 covered by CAIR, but removing Connecticut, Delaware, and Massachusetts from the CAIR group) and the District of Columbia. The D.C. Circuit's decision regarding the May 2005 CAIR, the time required to complete an alternative, and the concerns with and pending state challenges to the final Cross-State Rule have had a disruptive effect on certain aspects of initial implementation of the 2006 revised $PM_{2.5}$ NAAQS. The implications of the Cross-State Rule generated considerable interest among states, industry, and other stakeholders, and have been the topic of further debate during the 112th Congress.[35]

POTENTIAL IMPACTS OF THE 2006 PARTICULATES NAAQS

As discussed above, in setting and revising the NAAQS, the CAA directs the EPA Administrator to protect public health *with an adequate margin of safety*. This language has been interpreted, both by the agency and by the courts, as requiring standards based on a review of the health impacts, without consideration of the costs, technological feasibility, or other non-health criteria.[36] Costs and feasibility are generally taken into account in NAAQS implementation (a process that is primarily a state responsibility).

Nevertheless, EPA released a regulatory impact analysis (RIA) on October 6, 2006,[37] to meet its obligations under Executive Order 12866 and in compliance with guidance from the White House Office of Management and Budget.[38]

The RIA only analyzed the benefits and costs of implementing the $PM_{2.5}$ NAAQS. Citing time, data, and modeling limitations, EPA did not analyze the benefits and costs of retaining the PM_{10} standard.

EPA emphasized that the October 2006 RIA differed from typical RIAs in that it did not analyze the regulatory impact of a specific action, and that it is primarily for illustrative purposes.

The basis for the benefits calculations are reductions in ambient concentrations of $PM_{2.5}$ resulting from a reasonable, but speculative, array of cost-effective state implementation strategies selected by EPA for purposes of analysis. The analysis does not model the specific actions that each state will undertake in implementing the 2006 $PM_{2.5}$ NAAQS. EPA includes a detailed discussion of the limitations and uncertainties associated with the analysis.

In its impact analysis, EPA's estimated baseline case incorporated expected impacts of other national regulations that contribute to lowering $PM_{2.5}$ concentrations in future years. These other regulations included those addressing emissions from the power generation sector, in particular the 2005 Clean Air Interstate Rule (CAIR).[39] CAIR was expected to serve as the primary tool to assist downwind states in meeting the PM2.5 (and 8-hour ozone) NAAQS by mitigating interstate transport of sulfur dioxide (SO2) and nitrogen oxide (NOx) emissions from electric generating units that contribute to the formation of $PM_{2.5}$.[40] As discussed earlier in this report, the US Court of Appeals for the D.C. Circuit initially vacated CAIR in its entirety in a July 11, 2008,[41] decision, then modified its decision, allowing CAIR to remain in effect until a new rule is promulgated by EPA.[42] EPA subsequently published a final rule replacing CAIR, the Cross-State Air Pollution Rule, on August 8, 2011.[43]

The Cross-State Rule leaves the CAIR Phase 1 limits in place and sets new limits replacing CAIR's second phase in 2012 and 2014, up to three years earlier than CAIR would have. Building on emission reductions attributed to CAIR phase I, the Cross-State Rule is expected to result in relatively comparable benefits to those associated with CAIR in EPA's October 2006 RIA.

EPA's Monetized Benefits and Cost Estimates[44]

EPA estimated incremental costs of attaining the revised $PM_{2.5}$ standard based on a set of assumptions and extrapolations regarding currently designated nonattainment areas, likely control strategies and technologies and their associated engineering costs, emissions inventories and sources, and regional variability. EPA emphasized that the technologies and control strategies selected for analysis only illustrate one way for nonattainment areas to reach attainment, and that states will compile and evaluate a variety of programs and adopt those attainment strategies best suited for their specific local conditions. For purposes of comparing costs with monetized benefits, EPA estimated that the total annual mean social cost of attainment of the revised $PM_{2.5}$ NAAQS incremental to attainment of the 1997 standards would be $5.4 billion in 2020.

EPA's estimates of the monetized benefits of complying with the revised $PM_{2.5}$ standard reflect the valuation associated with predicted reductions in the incidence of certain health and social welfare effects. In the RIA, EPA presented a variety of benefits estimates based on morbidity impacts derived from several published epidemiological studies,[45] and mortality impacts derived from an American Cancer Society (ACS) Study[46] used in previous RIAs, the Harvard Six Cities Study,[47] and an expert elicitation study conducted by EPA in 2006.[48] *Table 2* below presents a range of EPA's estimated economic costs, monetized benefits, and net benefits associated with achieving the revised 2006 $PM_{2.5}$ NAAQS.

EPA estimated the total annual monetized benefits of attaining the revised $PM_{2.5}$ NAAQS would range from $15 billion to $17 billion based on the mortality function from the ACS study and morbidity function from the published epidemiology studies. Using the mortality function developed from the expert elicitation in conjunction with the morbidity function from the published studies, EPA's total annual benefits are estimated to range from $8 billion to $76 billion in 2020. EPA's estimated monetized benefits for 2020, like the cost estimates, were based on EPA's projected compliance schedule and are incremental to compliance with the 1997 $PM_{2.5}$ NAAQS by 2015.

According to the October 2006 RIA the estimated total annual health and welfare net benefits (subtracting social costs from the monetized benefits) in 2020 of attaining the 2006 $PM_{2.5}$ NAAQS ranged from $9 billion to $12 billion based on modeling of morbidity and mortality using published epidemiology studies. The benefits ranged from $2.4 billion to $70 billion, based on derivation mortality from expert elicitation and morbidity from the epidemiology studies. The ranges of estimates are forecasted using discount rates of 3% and 7%.

Table 2. EPA's Estimated Total Annual Monetized Benefits and Costs of Attaining Alternative $PM_{2.5}$ NAAQS in 2020 (1999 $ in billions)

		2006 $PM_{2.5}$ NAAQS (15/35 $\mu g/m^3$)			
	Discount Rate	Full Benefits	Cost	Net Benefits[a]	
Benefits based on American Cancer Society Study Mortality Function and Published Scientific Literature Morbidity Functions					
	3%	$17 confidence interval ($4 - $36)	$5.4	$12	
	7%	$14.5 confidence interval($3.5 - $31)	$5.4	$9	
Benefits Range based on Expert Elicitation Derived Mortality Function and Published Scientific Literature Morbidity Functions					
	Low Mean	High Mean		Low Mean	High Mean
	$9	$76			
3%	lower bound ($0.8 - $42)	upper bound ($19 - $150)	$5.4	$3.5	$70
	$8	$64			
7%	lower bound ($0.8 - $36)	upper bound ($16 - $130)	$5.4	$2.4	$59

Source: Prepared by the Congressional Research Service using data from the Environmental Protection Agency's Regulatory Impact Analysis (RIA) of the 2006 National Ambient Air Quality Standards for Fine Particle Pollution ($PM_{2.5}$); see Table ES-1, p. ES-7; Table ES-3, p. ES-9; Table 5-1, p. 5-3, Table 9-1, p. 9-2. http://www.epa.gov/ttn/ ecas/ria.html. Estimates and results are as reported by EPA and have been rounded after calculation.

Note: Results are rounded to two significant digits after calculation for presentation and computation as reported by EPA. Estimates (costs and benefits) reflect full attainment in 2020, which includes implementation of several national programs and are incremental to compliance with the 1997 $PM_{2.5}$ NAAQS. The discount rates are as recommended in EPA's Guidelines for Preparing Economic Analyses (2000) and OMB Circular A-4 (2003).

[a] For purposes of calculating net benefits, EPA uses the total social cost estimate, which is slightly higher than the engineering cost.

EPA's benefits and cost estimates were in terms of 1999 dollars and were incremental to the agency's modeled attainment strategy for the 1997 $PM_{2.5}$ NAAQS by 2015. The baseline case incorporated expected impacts associated with implementation of national regulations addressing emissions from the power generation sector, including the CAIR,[49] as well as regulation of various mobile sources, that contribute to lowering $PM_{2.5}$ concentrations in future years.

In addition to the monetized health benefits estimates, EPA estimated the monetary benefits associated with improvements in visibility in selected Class I national parks and wilderness areas.[50] EPA primarily used a stated preference approach which estimates values based on sampling surveys asking people what amount of compensation would be equivalent to a defined improvement in environmental quality. Extrapolating the results of a study based on a 1988 survey on recreational visibility value, EPA estimated visibility "willingness to pay" benefits to be $530 million in 2020 with attainment of the 2006 $PM_{2.5}$ NAAQS.[51]

EPA estimated the cost and benefits of a more stringent alternative $PM_{2.5}$ for purposes of comparative analysis. The comparative results are discussed in the "Concerns and Issues" section of this report.

Potential Health Impacts

The EPA health effects estimates were a primary component of its derivations of the monetized benefits discussed above. For purposes of illustration, *Table 3* summarizes EPA's predicted reductions in the incidence of a range of adverse health effects annually in 2020 for the 2006 $PM_{2.5}$ NAAQS, as reported in its RIA. The range of the estimated mean number of reductions in premature deaths is based on EPA's derivations using the ACS and the Harvard Six-Cities studies. EPA's mean estimates for the remaining adverse health effects were based on various epidemiology studies.[52]

EPA also estimated reductions in premature mortality based on the expert elicitation approach discussed above. The estimates were variable from expert to expert, ranging from a mean of 1,200 to 13,000 avoided premature deaths annually in 2020 resulting from attainment of the 2006 standards (15/35 $\mu g/m^3$) incremental to EPA's baseline strategy for the 1997 $PM_{2.5}$ NAAQS (15/65 $\mu g/m^3$).

When promulgating the 1997 $PM_{2.5}$ NAAQS, EPA estimated that compliance would result in the annual prevention of 15,000 premature deaths, 75,000 cases of chronic bronchitis, and 10,000 hospital admissions for respiratory and cardiovascular disease, as well as other benefits. These estimates have been the subject of significant debate and re-analysis. Since 1998, with dedicated funding from Congress, EPA accelerated its research and re-analysis on $PM_{2.5}$ to better understand the potential associated health effects and to develop ways to reduce risks.[53] The funding supported EPA intramural and extramural PM research projects and the establishment of five university-based PM research centers around the country.

Table 3. EPA's Predicted Reductions in Adverse Health Effects Annually in 2020 Associated with Meeting the 2006 $PM_{2.5}$ NAAQS

Adverse Health Effect	Predicted Reductionsa (estimated mean)
Premature deaths in individuals with preexisting cardiovascular and respiratory disease	2,500 to 5,700b
Cases of chronic bronchitis (age >25)	2,600
Cases of acute bronchitis (age 8-12)	7,300
Nonfatal heart attacks (age >71)	5,000
Hospital admissions for cardiovascular or respiratory symptoms (age >17)	1,630
Emergency room visits for asthma (age <19)	1,200
Cases of aggravated asthma (asthmatics age 6-18)	51,000
Cases of upper and lower respiratory symptoms (asthmatics age 6-18)	97,000
Days when individuals miss work (age 18-75)	350,000
Days when individuals must restrict their activities because of symptoms related to particle pollution (age 18-65)	2,000,000

Source: Prepared by the Congressional Research Service with data based on mortality and epidemiology studies presented in Chapter 5 of the Environmental Protection Agency Regulatory Impact Analysis of the 2006 National Ambient Air Quality Standards for Fine Particle Pollution ($PM_{2.5}$) Oct. 6, 2006, and available on EPA's website at http://www.epa.gov/ttn/ecas/ria.html. Estimates are rounded by EPA.

[a] For consistency with the emissions and benefits modeling, EPA used national population estimates based on the US Census Bureau projections. US Bureau of Census, 2000. Population Projections of the United States by Age, Sex, Race, Hispanic Origin and Nativity: 1999 to 2100. Population Projections Program, Population Division, US Census Bureau, available at http://www.census.gov/ population/projections/nation/ summary/np-t.txt.

[b] The range of premature death estimates reflects the mean estimates derived from the American Cancer Society study and the Harvard Six-Cities Study, respectively.

According to the February 2010 EPA report entitled *Our Nation's Air: Status and Trends Through 2008*,[54] nationally, annual and 24-hour $PM_{2.5}$

concentrations were 17% and 19% lower, respectively, in 2008 compared to 2001.

Based on a comparison of results from $PM_{2.5}$ monitoring locations (565 sites) for two three-year periods, 2001-2003 and 2006-2008, EPA reported that almost all sites showed a decline or little change in $PM_{2.5}$ concentrations.

Of the 565 monitoring sites, EPA reported that 16 sites in California, Illinois, Indiana, Michigan, Ohio, Utah, and West Virginia showed the greatest decreases in annual $PM_{2.5}$ concentrations, and five sites in California, Montana, Oregon, Pennsylvania, and Utah showed the greatest decrease in 24-hour $PM_{2.5}$ concentrations.

Four monitoring sites located in Montana, Arizona, and Wisconsin showed the greatest increase in annual $PM_{2.5}$ concentrations, but none were above the annual $PM_{2.5}$ NAAQS standard for the most recent three-year period (2006-2008).

Nineteen sites located throughout all regions of the United States showed an increase in 24-hour $PM_{2.5}$ concentrations. Seven of these sites are located in or near metropolitan areas in Arizona, California, Georgia, Montana, Virginia, and Washington were above the 24-hour $PM_{2.5}$ NAAQS standard for the most recent three-year period.

Most of the areas that showed the greatest improvement were the ones that generally had the highest concentrations in the earlier years. Despite the reported overall progress, in 2008 nearly 37 million people lived in counties with measured concentrations exceeding the annual and the 24-hour $PM_{2.5}$ national air quality standards, according to the EPA trends report.[55]

EPA's review leading up to the promulgation of the 2006 particulates NAAQS increased its confidence in earlier findings associating exposure to $PM_{2.5}$ with increases in respiratory health problems, hospitalizations for heart and lung disease, and premature death, particularly for children, the elderly, and those with preexisting heart and lung disease.[56]

However, as referenced earlier, findings and policy options presented for consideration in EPA's staff assessments,[57] and associated recommendations by the Clean Air Scientific Advisory Committee (CASAC)[58] in support of the ongoing next round of the agency's periodic review of the particulates NAAQS, have raised the question of whether the current standards provide an adequate margin of safety for protecting public health as required under the CAA.

The final EPA staff policy assessment released in April 2011 concluded that the evidence reviewed and analyzed generally supported consideration of tightening the $PM_{2.5}$ standards, and that it would be appropriate to consider retaining or revising the PM_{10} standard depending on the weight placed on the evidence and associated uncertainties.

CONCERNS AND ISSUES ASSOCIATED WITH THE REVISIONS OF THE PARTICULATES NAAQS

Congress and a wide variety of stakeholders closely followed the development of the 2006 particulates NAAQS since EPA's review began nearly 10 years ago. Congressional hearings were held regarding implementation and review of the particulates NAAQS.[59] Well before EPA formally proposed revising the particulates NAAQS, stakeholders were providing evidence and arguments at public hearings and other forums for their preferred recommendations. In general, business and industry opposed more stringent standards, and public health and environmental interest groups advocated tighter standards. EPA received thousands of comments during various stages of development of the particulates criteria document and in response to drafts of the EPA particulates staff paper. The agency reported receiving more than 120,000 comments in response to the January 2006 particulates NAAQS proposal.

The Administrator's proposed and final decisions represented the first time in the nearly 30-year history of the Clean Air Scientific Advisory Committee (CASAC) that the promulgated standards fell outside of the range of the scientific panel's recommendations.[60] In letters dated March 21, 2006, and September 29, 2006, the CASAC raised its concerns and objections regarding both PM_{10} and $PM_{2.5}$ proposed standards.[61]

The Administrator is not required by statute to follow CASAC's recommendations; the CAA (in Section 307(d)(3)) requires only that the Administrator set forth any pertinent findings, recommendations, and comments by CASAC and the National Academy of Sciences, and, if the Administrator's proposal differs in an important respect from any of their recommendations, provide an explanation of the reasons for such differences. Courts, in reviewing EPA regulations, also generally defer to the Administrator's judgment on scientific matters, focusing more on issues of procedure, jurisdiction, and standing. Nevertheless, CASAC's detailed objections to the Administrator's decisions and its description of the process as having failed to meet statutory and procedural requirements could play a role in the debate regarding EPA's reviews of NAAQS.

Petitions Challenging the 2006 PM NAAQS and the D.C. Circuit's February 24, 2009, Decision[62]

In December 2006, a diverse mix of petitioners, including states and state agencies,[63] environmental and public health organizations, and groups representing various industry and agriculture interests, petitioned the US Court of Appeals for the D.C. Circuit to review EPA's 2006 final rule revising the PM NAAQS. The court consolidated the cases.[64]

The petitioners, who challenged aspects of both the fine and coarse PM standards, were joined by various interveners and amici. Decisions regarding $PM_{2.5}$ standard focused primarily on the annual standard. Although the court's February 24, 2009, decision remanded aspects of EPA's decisions regarding the primary annual NAAQS for $PM_{2.5}$ to the agency for reconsideration, it did not vacate the annual standard, stating: "First, the EPA's failure adequately to explain itself is in principle a curable defect. Second, vacating a standard because it may be insufficiently protective would sacrifice such protection as it now provides, making the best an enemy of the good."[65]

Although aspects of the 2006 PM standard remanded to EPA for reconsideration by the court remain in effect and EPA's tightening of the 24-hour $PM_{2.5}$ standard was not affected, EPA's ongoing next round of periodic review of the standards, in part in response to the court's decision, has impacted the implementation process. EPA's eventual actions may have longer-term implications, particularly if the agency further tightens the primary or secondary annual standards. Although speculative at best, such actions could require modifications to SIPs submitted by existing previously designated nonattainment areas.

The following sections summarize key areas of concern following the promulgation of the 2006 PM NAAQS and include brief discussions of the D.C. Circuit's February 24, 2009, decision as it pertains to the particular issue area.

Fine Particulate ($PM_{2.5}$) Primary (Health) Standards

The final 2006 revised $PM_{2.5}$ NAAQS, were not as stringent as the levels recommended by the independent CASAC and those recommended by EPA professional staff, as noted above. EPA staff and CASAC recommendations for $PM_{2.5}$ included a range of levels more stringent than those proposed in January and finalized in September of 2006. In particular, the majority of the

CASAC panel "did not endorse the option of keeping the annual standard at its present value."[66] According to the CASAC:

> Of the options presented by EPA staff for lowering the level of the PM standard, based on the above considerations and the predicted reductions in health impacts derived from the risk analyses, most Panel members favored the option of setting a 24-hour PM2.5 NAAQS at concentrations in the range of 35 to 30 µg/m3 with the 98th percentile form, in concert with an annual NAAQS in the range of 14 to 13 µg/m3.67

Table 4 below compares the CASAC and EPA staff recommendations for $PM_{2.5}$ primary standards, the 1997 standards, and 2006 standards as proposed and promulgated.

Table 4. $PM_{2.5}$ Primary (Health) NAAQS: Final (2006), Proposed and Alternatives, and as Promulgated in 1997

$PM_{2.5}$ NAAQS Options	24-hour Primary(98th percentile)	Annual Primary (arithmetic mean)
1997 NAAQS	65 µg/m3	15 µg/m3
EPA staff paper (December 2005)	mid to lower range of 35-25 µg/m3	15 µg/m3
	or	
	mid to lower range of 40-30 µg/m3	14-12 µg/m3
CASAC (December 2005)	35-30 µg/m3	14-13 µg/m3
EPA Proposed Rule (January 2006)	35 µg/m3	15 µg/m3
EPA Final Rule (October 2006)	35 µg/m3	15 µg/m3

Source: Prepared by the Congressional Research Service (CRS), with information from EPA's final 2006 particulates NAAQS (71 Federal Register 61143-61233, Oct. 17, 2006), EPA's proposed particulates NAAQS (71 Federal Register 2620, Dec. 20, 2005), and related technical documents,68 available at http://www.epa.gov/air/particles/actions.html.

In response to the discrepancies between the proposal and the CASAC recommendations, then EPA Administrator Stephen Johnson indicated that his decision required consideration of a number of factors and "judgment based upon an interpretation of the evidence." The Administrator relied on the evidence of long-term exposure studies as the principal basis for retaining the annual $PM_{2.5}$ standard.[69] CASAC strongly disagreed with the Administrator's decision regarding the $PM_{2.5}$ annual standard and took the unprecedented step of urging reconsideration of the proposal.[70]

Many public comments received on EPA's proposed revisions to the $PM_{2.5}$ standards, most frequently from environmental and public health organizations, medical doctors and researchers, and the association representing state air quality regulators,[71] argued for standards as stringent or more stringent than those recommended by CASAC. In contrast, another group of commenters, generally representing industry associations and businesses, opposed revising the 1997 $PM_{2.5}$ standards, in some cases highlighting different aspects of the same research cited by the CASAC and others supporting tighter standards.[72] Some who opposed more stringent particulates NAAQS called attention to more recent studies of health effects attributable to particulates that demonstrate risk estimates are lower and less statistically significant than they were in 1997, when the last standard was set.[73]

In Section II of the preamble of the final October 2006 revisions, "Rationale for Final Decisions on Primary $PM_{2.5}$ Standards," EPA discussed its final decision with respect to the CASAC recommendations regarding the $PM_{2.5}$ annual standard. The Administrator differed with the CASAC with regard to the level of uncertainty associated with the agency's quantitative risk assessment and whether the results appropriately served as a primary basis for a decision on the level of the annual $PM_{2.5}$ standard. The Administrator further stressed the emphasis placed on the long-term means of the levels associated with mortality effects in the two key long-term studies[74] in determining the level of the annual standard. CASAC considered the evidence from specific short-term exposure studies as part of the basis for its recommendation for a lower annual standard level. As noted above, the CASAC expressed its objections to EPA's final 2006 particulates NAAQS in its September 29, 2006, letter to Administrator Johnson.[75]

With regard to $PM_{2.5,}$ the letter stated: "CASAC is concerned that the EPA did not accept our finding that the annual $PM_{2.5}$ standard was not protective of human health and did not follow our recommendation for a change in that standard."[76] The letter noted that "there is clear and convincing scientific evidence that significant adverse human-health effects occur in response to short-term and chronic particulate matter exposures at and below 15 $\mu g/m^3$," and noted that 20 of the 22 Particulate Matter Review Panel members, including all seven members of the statutory committee, were in "complete agreement" regarding the recommended reduction. "*It is the CASAC's consensus scientific opinion that the decision to retain without change the annual $PM_{2.5}$ standard does not provide an adequate margin of safety ... 'requisite to protect the public health' (as required by the Clean Air Act).*"[77] (Italics in original.)

Petitions from several states and state agencies, as well as the American Lung Association, Environmental Defense, and the National Parks Conservation Association, challenged EPA's decision not to strengthen the primary annual NAAQS standard for $PM_{2.5}$ ($15\mu g/m^3$). Their challenges were based, in part, on the agency's reliance primarily on long-term exposures and lack of consideration of vulnerable populations. The D.C. Circuit concluded that EPA failed adequately to explain why, in view of the risks posed by short-term exposures and the evidence of morbidity resulting from long-term exposures, its annual standard of $15\mu g/m^3$ is "requisite to protect health [with] an adequate margin of safety," including the health of vulnerable subpopulations.[78] In its February 24, 2009, decision, the court granted this part of the petition and remanded the standard to EPA for reconsideration, but did not vacate the standard. The court denied the same petitions with respect to review of the EPA analysis of long-term mortality studies, and its decision not to rely upon the quantitative risk assessment.

Potential Health Benefits of a More Stringent $PM_{2.5}$ Standard

In its RIA, EPA estimated the nationwide monetized human health and welfare benefits of attaining two suites of $PM_{2.5}$ NAAQS: (1) the revised $PM_{2.5}$ NAAQS, which include the 35 $\mu g/m^3$ daily (24-hour) standard and the unchanged 15 $\mu g/m^3$ annual standard, and (2) an alternative standard similar to the least stringent of the CASAC recommendations that includes a tighter annual standard of 14 $\mu g/m^3$ and the same 35 $\mu g/m^3$ daily (24-hour) standard. As discussed previously, EPA presented a variety of benefits estimates based on several epidemiological studies,[79] the American Cancer Society (ACS) Study[80] used in previous RIAs, the Harvard Six-Cities Study,[81] and an expert elicitation study conducted by EPA in 2006.[82]

EPA estimated that attainment of the more stringent alternative $PM_{2.5}$ NAAQS would result in $26 billion to $30 billion of total annual benefits in 2020, based on the ACS mortality function. This compares to a range of $15 billion to $17 billion estimated for compliance with the 2006 $PM_{2.5}$ NAAQS (see *Table 2* and discussion earlier in this report). EPA's estimate of annual benefits derived using the expert elicitation ranged from $15 billion to $140 billion for the more stringent alternative, compared to the expert estimates of

$8 billion to $76 billion for compliance with the 2006 standard. EPA also estimated the monetary benefits ("willingness to pay") associated with improvements in visibility in selected Class I national parks and wilderness areas would be $1.2 billion in 2020 with attainment of the more stringent alternative $PM_{2.5}$ standard analyzed, compared to $530 million with attainment of the revised $PM_{2.5}$ NAAQS.[83] EPA estimated the total annual cost associated with attainment of the alternative $PM_{2.5}$ NAAQS analyzed would be $7.9 billion in 2020, compared to $5.4 billion.

For purposes of illustration, *Table 5* below provides a comparison of EPA's predicted reductions annually for the 2006 $PM_{2.5}$ NAAQS (15/35 $\mu g/m^3$) with a more stringent alternative analyzed (14/35 $\mu g/m^3$), based on premature mortality estimates derived from the ACS and Harvard Six-Cities studies, and morbidity derived from various epidemiology studies. As discussed previously, a key component of EPA's monetized benefits estimates are the agency's predicted reductions in the incidence of premature deaths and a range of adverse health effects annually in 2020 associated with compliance of the 2006 $PM_{2.5}$ NAAQS.

As discussed earlier, EPA also used expert elicitation to characterize the concentration response function for the relationship between fine PM and premature mortality. The estimates EPA derived using this approach were variable from expert to expert, ranging from a mean of 2,200 to 24,000 avoided premature deaths annually in 2020 resulting from attainment of the more stringent alternative standard (14/35 $\mu g/m^3$) incremental to EPA's baseline strategy for the 1997 NAAQS (15/65 $\mu g/m^3$). For attainment of the 2006 standards (15/35 $\mu g/m^3$), EPA estimated 1,200 to 13,000 fewer premature deaths based on the expert elicitation.

EPA assessments[84] in support of the next round of periodic review of the particulates NAAQS have further stimulated the debate and generated considerable comment with regard to further tightening of both the annual and the 24-hour $PM_{2.5}$ standards. The assessments include findings that indicate that more stringent annual *and* 24-hour $PM_{2.5}$ standards could potentially reduce mortality risk from long-term exposure, as well as provide protection from high peak concentrations. In its review of EPA's draft assessments, members of the CASAC support the EPA staff's conclusion that "currently available information clearly calls into question the adequacy of the current standards" and that the staff's recommendation for consideration of a stricter $PM_{2.5}$ standard are "supported by the epidemiological and toxicological evidence, as well as by other risk and air quality information" compiled by EPA in related scientific assessments.[85]

Table 5. EPA's Predicted Reductions in Health Effects Annually in 2020 Associated with Meeting the 2006 $PM_{2.5}$ NAAQS and a More Stringent Alternative Annual Standard, Incremental to Attainment of the 1997 $PM_{2.5}$ Standards

	$PM_{2.5}$ NAAQS(15/35 μg/m3)	More Stringent Annual Std. (14/35 μg/m3)
Adverse Health Effect	Predicted Reductions[a] (estimated mean)	
Mortality		
Premature deaths in individuals with preexisting cardiovascular and respiratory disease		
Based on American Cancer Society and Six Cities Studies	2,500 to 5,700b	4,000 to 9,000[b]
Based on Expert Elicitation Results (low and high mean)	1,200 to 13,000	2,200 to 24,000
Morbidity		
Cases of chronic bronchitis (age >25)	2,600	4,600
Cases of acute bronchitis (age 8-12)	7,300	13,000
Nonfatal heart attacks (age >71)	5,000	8,700
Hospital admissions for cardiovascular or respiratory symptoms (age >17)	1,630	3,080
Emergency room visits for asthma (age <19)	1,200	3,200
Cases of aggravated asthma (asthmatics age 6-18)	51,000	79,000
Cases of upper and lower respiratory symptoms (asthmatics age 6-18)	97,000	153,000
Days when individuals miss work (age 18-75)	350,000	550,000
Days when individuals must restrict their activities because of symptoms related to particle pollution (age 18-65)	2,000,000	3,300,000

Source: Prepared by the Congressional Research Service with data presented in Chapter 5 of EPA's Agency Regulatory Impact Analysis of the 2006 National Ambient Air Quality Standards for Fine Particle Pollution ($PM_{2.5}$) Oct. 6, 2006, and available on EPA's website at http://www.epa.gov/ttn/ecas/ria.html. Estimates are as rounded by EPA.

[a] For consistency with the emissions and benefits modeling, EPA used national population estimates based on the US Census Bureau projections. US Bureau of Census. 2000. Population Projections of the United States by Age, Sex, Race, Hispanic Origin and Nativity: 1999 to 2100. Population Projections Program, Population Division, US Census Bureau, Available at http://www.census.gov/population/ projections/nation/ summary/np-t.txt.

[b] The range of reductions in premature deaths estimates reflect the mean estimates derived from the American Cancer Society study and the Harvard Six-Cities Study, respectively.

Coarse Particulate (PM_{10}) Primary Standards

Particle Size Indicator

EPA and most of the CASAC panel members concluded that there was a lack of evidence (often a lack of studies) on long-term adverse health effects of specific PM_{10} measurements to support the annual standard, and that there was a specific need to address particles ranging in size from 2.5 to 10 micrometers.[86] EPA's January 17, 2006, proposal would have replaced the existing particle size indicator of 10 micrometers (PM_{10}) with an indicator range of 10 to 2.5 micrometers ($PM_{10\text{-}2.5}$), referred to as inhalable (or thoracic) coarse particles, and setting a $PM_{10\text{-}2.5}$ *daily* standard of 70 $\mu g/m^3$ rather than the current PM_{10} daily standard of 150 $\mu g/m^3$. At the time of its proposal, EPA concluded that the scientific evidence supported the standard based on short-term exposure to certain coarse particles, particularly in urban and industrial areas.

In the final 2006 particulates NAAQS, EPA decided to maintain the PM_{10} indicator, citing the limited body of evidence on health effects associated with thoracic coarse particles from studies that use $PM_{10\text{-}2.5}$ measurements. The agency also determined that the only studies of clear quantitative relevance to health effects most likely associated with thoracic coarse particles used PM_{10} measurements. The 2006 particulates NAAQS retained the PM_{10} indicator and the *daily* (24-hour) standard of 150 $\mu g/m^3$.

In its September 29, 2006, letter, the CASAC said it was "completely surprised" at the decision to revert to the use of PM_{10} as the indicator for coarse particles, noting that the option of retaining the existing daily PM_{10} standard was not discussed during the advisory process and that CASAC views this decision as "highly-problematic since PM_{10} includes both fine and coarse particulate matter." The CASAC did agree that having a standard for PM_{10} was better than no standard.

Included in the October 17, 2006, *Federal Register* Notice, EPA promulgated a new federal reference method (FRM) for measurement of mass concentrations of $PM_{10\text{-}2.5}$ in the atmosphere as the standard of reference for measurements of $PM_{10\text{-}2.5}$ concentrations in ambient air. EPA anticipated that the new FRM should provide a basis for gathering scientific data to support future reviews of the particulates NAAQS.[87]

EPA's January 17, 2006, proposal to change the indicator of the standard for coarse particles was in response to a 1999 US Court of Appeals for the DC Circuit decision[88] directing EPA to ensure that the standard did not duplicate the regulation of fine particles. The agency contends that it addressed the concerns raised by the court regarding PM_{10} as an indicator for inhalable coarse particulate matter in its rationale in the final 2006 particulates NAAQS, announced September 21, 2006.[89]

In its February 24, 2009, decision,[90] the D.C. Circuit denied petitions by industry and agricultural groups for review of the primary daily standard for $PM_{10.}$ In their petition, the American Farm Bureau Federation, the National Pork Producers Council, the National Cattlemen's Beef Association, and the Agricultural Retailers Association challenged EPA's decision to retain a daily standard for PM_{10} including that in nonurban areas. They also challenged the decision to retain the indicator particle size indicator of 10 micrometers (PM_{10}), and the primary daily standard of 150 $\mu g/m^3$. The court denied the petitions, finding EPA's decision and explanation on all three issues to be reasonable.

Environmental groups petitioned for review of EPA's decision to revoke the primary annual standard for $PM_{10.}$ The D.C. Circuit denied the petition finding EPA's decision reasonable because the latest scientific evidence, as shown in the Criteria Document and the Staff Paper, do not indicate long-term exposure to coarse particles poses a health risk.

Rural PM_{10} Sources

In addition to the changes to the coarse particulates indicator, EPA had proposed focusing the $PM_{10-2.5}$ standard only on "urban and industrial" sources. Particles typical to rural areas including "*windblown dust and soils and PM generated by agricultural and mining sources*" would *not* have been subject to this standard. Additionally, EPA proposed revoking the current 24-hour PM_{10} standards, except in areas that have (1) monitors indicating violation, and (2) a population of 100,000 or more. The emphasis on urban and industrial areas in the January 2006 proposal was based on the findings reported in the Criteria Document, the PM staff paper, and the CASAC conclusion that "the evidence for the toxicity of $PM_{10-2.5}$ comes from studies conducted primarily in urban areas and is related, in large part, to the re-entrainment of urban and suburban road dusts, as well as primary combustion products."[91]

EPA's proposal to exclude any ambient mix of $PM_{10-2.5}$ that is dominated by rural windblown dust and soils and particulates generated by agricultural and mining sources, and how EPA would distinguish the sources during its implementation, raised a number of questions and resulted in numerous comments. In response to the proposal, in its March 21, 2006, letter to the EPA Administrator, the CASAC stated that while it had recognized the scarcity of information on the toxicity of rural dust, it "neither foresaw nor endorsed a standard that specifically exempts all agricultural and mining sources, and offers no protection against episodes of urban-industrial $PM_{10-2.5}$ in areas of populations less than 100,000." The committee strongly recommended "expansion of our knowledge of the toxicity of $PM_{10-2.5}$ dusts rather than exempting specific industries (e.g., mining, agriculture)."[92]

Several members of the House Committee on Agriculture submitted a letter to EPA Administrator Stephen Johnson in July 2006 conveying support for the agency to maintain its provision to exclude agriculture and mining dust and similar sources of coarse particulates in the particulates NAAQS, as had been proposed.[93] In the final rule, EPA indicated that with the exception of representatives of those sources that would have been excluded under the proposal (e.g., agriculture and mining), most commenters opposed the exclusion. Those opposed included environmental and public health groups, state and local agencies, and industries not excluded from the proposed indicator (e.g., transportation and construction).

EPA did not exclude any areas or types of particles in the final 2006 particulates NAAQS revisions, based on further consideration of the data and in response to comments. In its rationale for the final PM_{10} standard, EPA continued to acknowledge that there is far more evidence concerning health effects associated with thoracic coarse particles in urban areas than in non-urban areas. However, EPA also stated that "the existing evidence is inconclusive with regard to whether or not community-level exposures to thoracic coarse particles are associated with adverse health effects in non-urban areas."[94] EPA indicated that it would be expanding its research and monitoring programs to collect additional evidence on the differences between coarse particles typically found in urban areas and those typically found in rural areas. EPA published a final rule amending its national air quality monitoring requirements in a separate Notice in the October 17, 2006, *Federal Register*.[95] Under the revised monitoring strategy, EPA and the states are adding measurement of "inhalable coarse particles," (i.e. $PM_{10-2.5}$) at the 75 multi-pollutant monitoring sites added per the rule. Many of the existing monitors across the country sampling levels of PM_{10} remained in place, especially in areas that exceed the 24-hour standard. However, monitors in some states that are redundant or are measuring air quality concentrations well below the level of the 24-hour PM_{10} standard have been shut down.

In contrast to objections regarding other aspects of EPA's final 2006 particulates NAAQS revisions, the CASAC agreed with the EPA decision against including exemptions in its September 29, 2006, letter to the EPA

Administrator. However, a number of those representing agriculture interests, including some Members of Congress,[96] remained concerned that EPA's decision not to include the exclusions in the final 2006 particulates NAAQS will result in unnecessary burdens on the agricultural community. In its February 24, 2009, decision,[97] the D.C. Circuit found that EPA provided reasonable explanation for its decisions not to set separate urban and nonurban PM_{10}, rejecting the petitioners challenge to the level of PM_{10} standard.[98]

Because the PM_{10} standard was not strengthened no new areas, including rural or non urban areas, will be designated as nonattainment for PM_{10}. To the contrary, some counties previously designated nonattainment have been determined by EPA to be in attainment since the 2006 revisions to the particulates NAAQS. Additionally, states' nonattainment recommendations for the 2006 $PM_{2.5}$ NAAQS and EPA's August 2008 proposed modifications to these recommendations do not include new rural or non-urban areas. The map in *Figure 1* below shows the status of nonattainment of the PM_{10} NAAQS.

As depicted on the map, the majority of the United States is in attainment of the PM_{10} NAAQS (including Hawaii, which is not shown on the map). Since 1990, EPA designated 87 areas as nonattainment with the PM_{10} NAAQS.[99] As of July 2009, 40 of the original 87 PM_{10} areas have been redesignated to attainment. Of the remaining 47 nonattainment areas (40 counties with a population of 25.5 million), 26 areas are currently not meeting the PM_{10} standard or have insufficient data to determine attainment based on 2005-2007 data. The remaining counties have submitted the required SIPs for PM_{10} but have not yet been formally redesignated to attainment.[100]

On April 22, 2011, EPA released a final staff policy assessment in support of the ongoing PM review.[101] The assessment, which addresses both fine ($PM_{2.5}$) and coarse (PM_{10}) particulates, summarizes EPA staff review of the adequacy of the current standards and their analyses of the scientific evidence, quantitative assessments and air quality analyses in support of potential alternative policy options for consideration by the agency prior to rulemaking. With regard to coarse particulates, EPA staff concluded that it would be appropriate to consider retaining or revising the current 24-hour (daily) PM_{10} standard depending on the weight placed on the evidence and its associated uncertainties. Options for revising the standard range from setting a standard "generally equivalent" to the current standard that better targets protection, to lowering the current level of the *daily* standard from 150 micrograms per cubic meter ($\mu g/m^3$) to a range of 85 to 65 $\mu g/m^3$ combined with a change in the way EPA determines whether an area is meeting the standard (the "form").

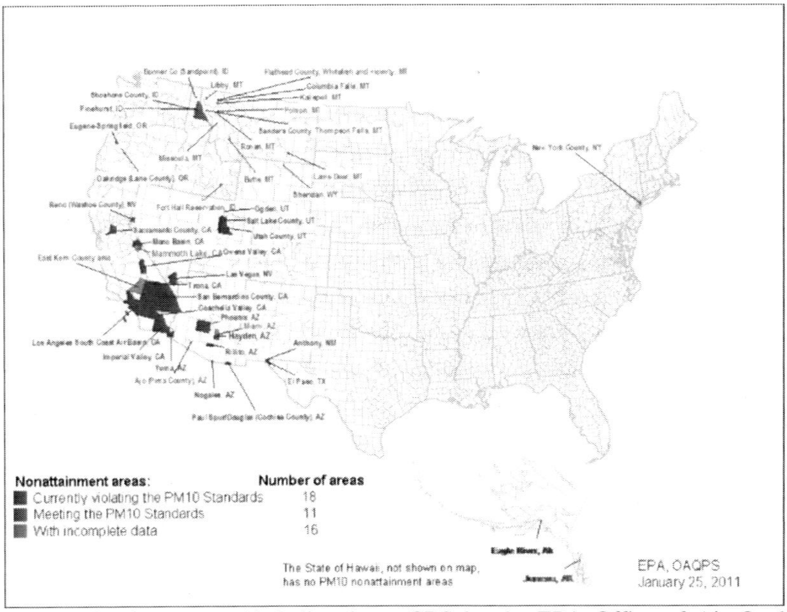

Source: Updated map provided directly to CRS by the EPA Office of Air Quality Planning and Standards (OAQPS), January 28, 2011. Areas not highlighted on the map are designated attainment/unclassifiable. There are no PM_{10} nonattainment areas in Hawaii, which was not included on the EPA map.

Figure 1. Status of Current PM_{10} Nonattainment Areas, Based on 2006-2008 Air Quality.

Previous draft EPA assessments[102] in support of the next round of periodic review of the particulates NAAQS stimulated debate and generated considerable comment, particularly within the agriculture community, with regard to potential further tightening the PM_{10} standards. With regard to the adequacy of the current primary 24-hour PM_{10} standard, EPA staff preliminary findings suggested that consideration could be given to retaining or revising the current standard downward in a June 2010 second external review draft policy assessment. In its August 2010 review of the draft assessment, CASAC recommended that "the primary standard be revised downwards and not retained," but did not agree that the available scientific evidence strongly supports the EPA staff proposed upper bound standard level of 85μg/m^3, favoring a more stringent level to "provide enhanced protection, somewhere in the range of 75 – 65 μg/m^3 …"[103]

In July and August 2010 letters to the EPA Administrator, some Members of Congress raised their concerns about EPA's July 2010 staff recommendations and the potential impacts on agriculture associated with tightening the PM_{10} standards, encouraging EPA to retain the current primary and secondary standards.[104]

Secondary $PM_{2.5}$ and PM_{10} Standards

The EPA proposal, and the final 2006 particulates NAAQS, set the secondary standard for PM_{10} and for $PM_{2.5}$ at the same level as their primary standard. The PM staff paper and the CASAC both recommended secondary standards at levels different from the primary in order to be more protective of visibility, and the CASAC reiterated the recommendations in its March 21, 2006, and September 29, 2006, letters to the EPA Administrator.

For $PM_{2.5,}$ the EPA PM staff paper and most of the CASAC panel recommended consideration of a sub-daily standard with a level in the range of 20 to 30 $\mu g/m^3$ for a four- to eight-hour midday time period, with a 92nd to 98th percentile form, as opposed to the primary daily standard at 35 $\mu g/m^3$, based on the current three-year average of the 98th percentile of 24-hour $PM_{2.5}$ concentrations. Although the CASAC ultimately agreed with setting a secondary standard at the same level as the primary standard based on the coarse particulates indicator $PM_{10-2.5,}$ the committee recommended that the standard not be limited to urban areas, as EPA had proposed.

In its February 24, 2009, decision,[105] the D.C. Circuit granted in full a petition from environmental groups challenging EPA's decision to set the $PM_{2.5}$ secondary NAAQS for protecting public welfare at the same level as the primary NAAQS for protecting public health. The decision remanded the $PM_{2.5}$ secondary NAAQS to EPA for reconsideration. The court found that EPA unreasonably concluded that the NAAQS are adequate to protect public welfare from adverse effects on visibility.

In assessments in support of the next round of the agency's periodic review of the particulate NAAQS referenced earlier in this report, EPA staff recommends revising the secondary standards for $PM_{2.5}$ based on preliminary review of data.[106]

Exclusion of More Recent Research

A number of stakeholders commented that EPA should have considered certain studies that were published too recently to have been included in the

2004 criteria document that, they argued, increased the uncertainty about possible health risks associated with exposure to particulates. Others contended that there are new studies (some of them the same studies) in support of their arguments for a lower (more stringent) level to protect health. Some who opposed more stringent standards commented that the agency should have delayed its decision regarding the PM NAAQS to take into consideration several of these studies.

At the time of the proposal, EPA declared its intention to review and evaluate significant new studies developed since 2002, and those published since the close of the criteria document, during the comment period.[107] With the release of its final 2006 particulates NAAQS, EPA acknowledged that these studies provided expansion of the science and some insights regarding particulates exposure and related health effects, but determined that the new data "do not materially change any of the broad scientific conclusions regarding the health effects of PM exposure made in the 2004 PM Air Quality Criteria Document."[108]

CONCLUSION

EPA's October 17, 2006, promulgation of the final modifications to the existing particulates NAAQS following completion of its statutorily required review sparked interest and conflicting concerns among a diverse array of stakeholders, and in Congress.

As a result of EPA's 2006 tightening of the fine particulates ($PM_{2.5}$) NAAQS, additional areas have been classified as nonattainment and need to implement new controls on particulate matter in order to comply. States and local governments will be required to develop and implement new or modified plans for addressing emissions in those areas that do not meet the 2006 standards. A stricter standard means increased costs for the transportation and industrial sectors most likely to be affected by particulate matter controls, including utilities, refineries, and the trucking industry. In terms of public health, a stricter standard is expected to result in fewer adverse health effects for the general population and particularly sensitive populations, such as children, asthmatics, and the elderly, with the estimated value of these benefits significantly outweighing the costs.

Because of health and cost implications, NAAQS decisions have been the source of significant concern to some in Congress. The evolution and development of the particulates NAAQS, in particular, have been the subject of extensive oversight. As a result, the D.C. Circuit's February 2009 decision regarding challenges to the 2006 PM NAAQS, delays in implementing the 2006 PM2.5 NAAQS resulting from the Administration's review of the final designations, and recent findings reported in assessments supporting EPA's next round of periodic review of the particulate NAAQS have prompted renewed interest in several of the issues associated with the promulgation of the 2006 PM NAAQS discussed throughout this report.

End Notes

[1] Sections 108-109 of the Clean Air Act.

[2] The NAAQS is for NO2; nitrogen gases that are ozone precursors are referred to as NOx.

[3] 71 *Federal Register* 61143-61233, October 17, 2006. See also EPA's PM Regulatory Actions website at http://epa.gov/pm/actions.html.

[4] 52 *Federal Register* 24634-24715, July 1, 1987.

[5] A micrometer is 1/1000th of a millimeter; there are 25,400 micrometers in an inch.

[6] 62 *Federal Register* 38652-38896, July 18, 1997. See CRS Report RL32431, *Particulate Matter (PM2.5): Implementation of the 1997 National Ambient Air Quality Standards (NAAQS)*, by Robert Esworthy.

[7] Section 109(d)(1) of the CAA. According to the statute, EPA is required to review the latest scientific studies and either reaffirm or modify the NAAQS every five years, however EPA has rarely met this requirement and in some cases has been forced to comply through litigation.

[8] For information regarding the CASAC PM review panel and its activities and reports, see http://yosemite.epa.gov/sab/ sabpeople.nsf/WebCommittees/CASAC.

[9] Section 109(d)(2) of the Clean Air Act.

[10] American Farm Bureau Federation v. US EPA, No. 06-1410 (D.C. Cir., February 24, 2009).

[11] Three amici states filed jointly with the 13 party states.

[12] Cases were consolidated with American Farm Bureau Federation v. US EPA, 2009 Westlaw 437050, No. 06-1410 (D.C. Cir. 2006).

[13] For a more detailed discussion of the NAAQS implementation process for designating geographical nonattainment areas with respect to the tightening of the PM2.5 standards under the 2006 particulates NAAQS, including comparisons with the final designations under the 1997 PM2.5 NAAQS, see CRS Report R40096, *2006 National Ambient Air Quality Standards (NAAQS) for Fine Particulate Matter (PM2.5): Designating Nonattainment Areas*, by Robert Esworthy.

[14] EPA had previously targeted proposing its decision regarding changes to the PM standards by February 2011 and October 2011 for final standards; US Environmental Protection Agency, "Federal Implementation Plans to Reduce Interstate Transport of Fine Particulate Matter and Ozone: Proposed Rule," 75 *Federal Register* 45220, August 2, 2010.

[15] US Environmental Protection Agency, "Release of Final Document Related to the Review of the National Ambient Air Quality Standards for Particulate Matter," 76 *Federal Register* 22665, April 22, 2011, http://www.epa.gov/ttn/naaqs/standards/pm/data/20110419 pmpafinal.pdf.

[16] See footnote 15, and also US Environmental Protection Agency, *Policy Assessment for the Review of the PM NAAQS—First External Review Draft*, EPA 452/P-10-003, March 2010; US Environmental Protection Agency, *Policy Assessment for the Review of the PM NAAQS—Second External Review Draft*, EPA-452/P-10-007, June 2010; US Environmental Protection Agency, *Integrated Science Assessment for Particulate Matter—Final Report*, EPA 600/R-08/139F, December 2009; and US Environmental Protection Agency, *Risk Assessment to Support the Review of the PM Primary National Ambient Air Quality Standards—External Review Draft*, EPA 450/P_09-006, September 2009. Assessments are available at http://www.epa.gov/ttn/naaqs/standards/pm/s_pm_index.html. US Environmental Protection Agency, "Release of Draft Documents Related to the Review of the National Ambient Air Quality Standards for Particulate Matter," 74 *Federal Register* 46589, September 10, 2009.

[17] EPA Clean Air Advisory Committee (CASAC) Particulate Matter (PM) Review Committee: *CASAC Review of Policy Assessment for the Review of the PM NAAQS—Second External Review Draft (June 2010)*, August 10, 2010, Draft, http://yosemite.epa.gov/sab/sabproduct.nsf/WebCASAC/AD891A65C35DC3738525777B005B0A6F/$File/PM+PA-2+letter+8-10-10.pdf; *CASAC Review of Policy Assessment for the Review of the PM NAAQS - First External Review Draft (March 2010)*, May 17, 2010, E504EE3276D87A9E8525772700647AFB/$File/EPA-CASAC-10-011- unsigned.pdf. See also CASAC's *Review of Quantitative Health Risk Assessment for Particulate Matter—Second External Review Draft, (February 2010)*, April 15, 2010, http://yosemite.epa.gov/sab/sabproduct.nsf/WebCASAC/ recentadditions.

[18] The use of public welfare in the CAA "includes, but is not limited to, effects on soils, water, crops, vegetation, manmade materials, animals, wildlife, weather, visibility, and climate, damage to and deterioration of property, and hazards to transportation, as well as effects on economic values and on personal comfort and well-being, whether caused by transformation, conversion, or combination with other air pollutants" (42 USC. 7602(h)).

[19] CASAC Particulate Matter (PM) review of EPA's *Review of the National Ambient Air Quality Standards for Particulate Matter: Policy Assessment of Scientific and Technical Information (Second Draft PM Staff Paper, January 2005*, EPA-SAB-CASAC-05-007, June 6, 2005), available on the EPA CASAC website; see "Final Reports by Fiscal Year" at http://yosemite.epa.gov/sab/sabproduct.nsf/WebReportsbyYearCASAC.

[20] Based on the findings in the EPA PM criteria document and staff paper, and the CASAC's concurrence, that the studies reviewed do not provide sufficient evidence regarding *long-term* exposure to warrant continuation of an annual standard, see 71 *Federal Register* 2653, *Section III. Rationale for Proposed Decision on Primary* $_{PM10}$ *Standards,* January 17, 2006.

[21] See CRS Report R40096, *2006 National Ambient Air Quality Standards (NAAQS) for Fine Particulate Matter (PM2.5): Designating Nonattainment Areas*, by Robert Esworthy.

[22] For a general overview of the NAAQS designations process, see EPA's "Designations" website at http://www.epa.gov/air/urbanair/designations.html.

[23] Section 110(a)(2) of the Clean Air Act. For a general overview of the NAAQS implementation plans process, see EPA's "State Implementation Plan Overview" website at http://www.epa.gov/air/urbanair/sipstatus/overview.html.

[24] 74 *Federal Register* 58688-58781; see also "Area Designations for 2006 24-Hour Fine Particulate (PM2.5) Standards—Regulatory Actions," http://www.epa.gov/pmdesignations/2006standards/regs.htm#4.

[25] For detailed $PM_{2.5}$ state/county geographical designation recommendations by EPA and those from individual states and tribes, for the 1997 and for the 2006 $PM_{2.5}$ NAAQS, see http://www.epa.gov/pmdesignations.

[26] In addition, in the February 3, 2011, *Federal Registe*r, EPA published its final designations of three areas as "nonattainment" or "unclassifiable/attainment" for the 2006 24-hour $PM_{2.5}$ NAAQS that were deferred at the time of the agency's November 13, 2009, promulgated designations. Designations were deferred so that EPA could evaluate measurements of high fine particle concentrations by violating monitors. EPA designated a portion of Pinal County, AZ, as nonattainment. Eight surrounding counties were designated unclassifiable/attainment, and Gila River Indian Community and the Ak-Chin Indian community designations were deferred pending completion of the formal consultation process. Plumas County, CA, and Shasta County, CA, as well as surrounding counties, were designated unclassifiable/attainment.

[27] The review of these regulations was initiated, in part, in response to a White House January 20, 2009, memorandum, Memorandum to Heads and Acting Heads of Executive Departments and Agencies from Rahm Emanuel, Assistant to the President and Chief of Staff, regarding Regulatory Review, January 20, 2009, 74 *Federal Register* 4435, January 26, 2009; the memorandum is also available at http://www.whitehouse.gov/sites/ default/files/omb/assets/ information_and_regulatory_affairs/regulatory_review_012009.pdf. Also see Office of Management and Budget website, "Regulatory Matters" - OIRA Policies and Practices, at http://www.whitehouse.gov/omb/ international_regulatory_cooperation, and the Office of Management and Budget's subsequent January 21, 2009, memorandum regarding regulatory review, Memorandum to Heads and Acting Heads of Executive Departments and Agencies from Peter R. Orszag, Director, Office of Management and Budget, regarding Implementation of Memorandum Concerning Regulatory Review, January 21, 2009; see Office of Management and Budget website "Regulatory Matters" - OIRA Policies and Practices, at http://www.whitehouse.gov/omb/ international_regulatory_cooperation.

[28] Section 110(a)(2) of the Clean Air Act. For a general overview of the NAAQS implementation plans process, see EPA's "State Implementation Plan Overview" at http://www.epa. gov/air/urbanair/sipstatus/overview.html.

[29] Section 172 of the Clean Air Act. See EPA's "State Implementation Plan Overview" at http://www.epa.gov/air/ urbanair/sipstatus/overview.html.

[30] Under section 172(a)(2)(A) of the CAA, EPA may grant an area an extension of the initial attainment date for one to five years (in no case later than 10 years after the designation date for the area). A state requesting an extension must submit an implementation plan (SIP) by the required deadline that includes, among other things, sufficient information demonstrating that attainment by the initial attainment date is "impracticable."

[31] 64 *Federal Register* 35714-35774, July 1, 1999. See CRS Report RL32483, *Visibility, Regional Haze, and the Clean Air Act: Status of Implementation*, by Larry Parker and John Blodgett, also CRS Report RL32927, *Clean Air Interstate Rule: Review and Analysis*, by Larry Parker.

[32] Promulgated under the CAA, 42 USC. 7401 et seq., see US EPA, "Ambient air quality standards, national—Fine particulate matter and ozone; interstate transport control measures," 70 *Federal Register* 25162, May 12, 2005.

[33] North Carolina v. EPA 531 F.3d 896 (D.C. Cir. 2008), see CRS Report RL34589, *Clean Air After the CAIR Decision: Multi-Pollutant Approaches to Controlling Powerplant Emissions*, by James E. McCarthy, Larry Parker, and Robert Meltz; also see EPA's Clean Air Interstate Rule website at http://www.epa.gov/CAIR/index.html.

[34] See US EPA, "Federal Implementation Plans: Interstate Transport of Fine Particulate Matter and Ozone and Correction of SIP Approvals," 76 *Federal Register* 48208-48483, August 8, 2011, http://www.gpo.gov/fdsys/pkg/FR2011-08-08/pdf/2011-17600.pdf. EPA announced the release of the final rule on July 6, 2011. Explanatory and background material can be found on EPA's website at http://www.epa.gov/crossstaterule/actions.html. When proposed in August 2010, the Cross-State Rule was referred to as the Clean Air Transport Rule. The name change to "Cross-State Rule" occurred late in the development of the final rule. As a result, many of the explanatory materials, including the final Regulatory Impact Analysis, refer to the "Transport Rule."

[35] For example, see House Committee on Science, Space, and Technology, full-committee hearing, "Out of Thin Air: EPA's Cross-State Air Pollution Rule," September 15, 2011, http://science.house.gov/hearing/full-committee-hearing- epas-cross-state-air-pollution-rule.

[36] With regard to the non-relevance of cost considerations, see generally Whitman v. American Trucking Associations, 531 US 457, 465-472, 475-76 (2001).

[37] EPA's Regulatory Impact Analysis (RIA) of the 2006 National Ambient Air Quality Standards for Fine Particle Pollution (PM2.5), available at http://www.epa. gov/ttn/ecas/ ria.html.

[38] 58 *Federal Register* 51735, October 4, 1993. See the White House OMB website, *Regulatory Matters*, at http://www.whitehouse.gov/omb/regulatory_affairs/default.

[39] CAA, 42 USC. 7401 et seq., 70 *Federal Register* 25162, May 12, 2005.

[40] Sulfur dioxide (SO2) is a precursor (a pollutant that is transformed in air to form another air pollutant) contributing to the formation of PM2.5 concentrations, and NOx is a precursor contributing to the formation of both ozone and PM2.5 concentrations. EPA has concluded that SO2 and NOx emissions, through the phenomenon of air pollution transport, contribute significantly to downwind nonattainment, or interfere with maintenance, of the PM2.5 and 8-hour ozone NAAQS (70 *Federal Register* 25162, May 12, 2005).

[41] 531 F.3d 896 (D.C. Cir. 2008). See CRS Report RL34589, *Clean Air After the CAIR Decision: Multi-Pollutant Approaches to Controlling Powerplant Emissions*, by James E. McCarthy, Larry Parker, and Robert Meltz.

[42] 550 F.3d 1176 (D.C. Cir. 2008).

[43] See US EPA, "Federal Implementation Plans: Interstate Transport of Fine Particulate Matter and Ozone and Correction of SIP Approvals," 76 *Federal Register* 48208-48483, August 8, 2011, http://www.gpo.gov/fdsys/pkg/FR2011-08-08/pdf/2011-17600.pdf. EPA announced the release of the final rule on July 6, 2011. Explanatory and background material can be found on EPA's website at http://www.epa.gov/crossstaterule/actions.html. When proposed in August 2010, the Cross-State Rule was referred to as the Clean Air Transport Rule. The name change to "Cross-State Rule" occurred late in the development of the final rule. As a result, many of the explanatory materials, including the final Regulatory Impact Analysis, refer to the "Transport Rule."

[44] US EPA, EPA's RIA of the 2006 NAAQS for Fine Particle Pollution ($PM_{2.5}$), available at http://www.epa.gov/ttn/ ecas/ria.html.

[45] The studies from which EPA health effect estimates are drawn are listed in Table 5-7 of the RIA for the 2006 PM2.5 NAAQS.

[46] Pope, C. Arden, III et al. "Particulate Air Pollution as a Predictor of Mortality in a Prospective Study of US Adults." *American Journal of Respiratory and Critical Care Medicine* 151 (1995): 669-674.

[47] Dockery, Douglas W. et al. "An Association Between Air Pollution and Mortality in Six US Cities." *New England Journal of Medicine* 329 (1993): 1753-1759. See also the Health Effects Institute, "Statement: Synopsis of the Particle Epidemiology Reanalysis Project."

Reanalysis of the Harvard Six Cities Study and the American Cancer Society Study of Particulate Air Pollution and Mortality, July 2000 (includes November 1, 2001 errata sheet), p. I, http://pubs.healtheffects.org/getfile.php?u=271.

[48] See Chapter 5 of EPA's October 6, 2006, RIA http://www.epa.gov/ttn/ecas/ria.html.

[49] 70 *Federal Register* 25162, May 12, 2005.

[50] Defined as areas of the country such as national parks, national wilderness areas, and national monuments that have been set aside under Section 162(a) of the Clean Air Act to receive the most stringent degree of air quality protection.

[51] See Appendix I in *Visibility Benefits Methodology* of EPA's October 6, 2006, RIA http:// www.epa.gov/ttn/ecas/ ria.html.

[52] See Table 5-7 in Chapter 5 of EPA's October 6, 2006, RIA: http:// www. epa.gov/ ttn/ecas/ ria.html.

[53] Congress increased EPA's appropriations for particulate matter research from $18.8 million in FY1997 (H.Rept. 104- 812) to $49.6 million in FY1998 (H.Rept. 105-297). PM research appropriations averaged more than $60 million per year from FY1999 through FY2004, and Congress provided $60.5 million for FY2005. Congress has not identified PM research funding since FY2005.

[54] EPA-454/R-09-002, February 2010 http://www.epa.gov/airtrends/2010/index.html.

[55] Refers to trends in the measured $PM_{2.5}$ concentrations relative to their air quality standards in those areas where monitors are located across the country, and does reflect all designated nonattainment areas. Comparatively, the combined population residing in EPA's final nonattainment designation areas for the 1997 $PM_{2.5}$ NAAQS, consisting of 208 counties in 20 states and the District of Columbia, is almost 90 million.

[56] EPA criteria and technical documents in support of the October 17, 2006, final particulates NAAQS, the December 20, 2005, proposal, and the 1997 NAAQS, are available at http:// www.epa.gov/ttn/naaqs/standards/pm/ s_pm_index.html.

[57] See footnote 15 and footnote 16.

[58] See footnote 17.

[59] US Senate Committee on Environment and Public Works, full Committee, *The Science and Risk Assessment Behind the EPA's Proposed Revisions to the Particulate Matter Air Quality Standards*, July 19, 2006; Subcommittee on Clean Air, Climate Change, and Nuclear Safety, *EPA's Proposed Revisions to the Particulate Matter Air Quality Standards*, July 13, 2006, and *Implementation of the Existing Particulate Matter and Ozone Air Quality Standards*, November 10, 2005.

[60] For a discussion of issues regarding the CASAC, see CRS Report RL33807, *Air Quality Standards and Sound Science: What Role for CASAC?*, by James E. McCarthy.

[61] Letter of Dr. Rogene Henderson, Chair, Clean Air Scientific Advisory Committee, to the Hon. Stephen Johnson, Administrator, US EPA, March 21, 2006, EPA-CASAC-LTR-06-002, available at http://yosemite.epa.gov/sab/ sabproduct.nsf/CD706C976DAC62B3 85257139 0081CC21/$File/casac-ltr-06-002.pdf; and letter of Dr. Rogene Henderson, Chair of the Clean Air Scientific Advisory Committee et al. to Hon. Stephen L. Johnson, EPA Administrator, September 29, 2006, EPA-CASAC-LTR-06-003, available at http://yosemite.epa.gov/sab/ sabproduct.nsf/1C69E987731CB7758525 71FC00499A10/ $File/casac-ltr-06-003.pdf.

[62] See footnote 10.

[63] New York, California, Connecticut, Delaware, Illinois, Maine, New Hampshire, New Jersey, New Mexico, Oregon, Pennsylvania, Rhode Island, Vermont), the District of Columbia, and the South Coast Air Quality Management District.

[64] See footnote 12.

[65] See footnote 10.

[66] CASAC PM Review Panel report, p. 7, June 2005, http://www.epa.gov/ttn /naaqs/ standards /pm/s_pm_index.html.

[67] See footnote 66.

[68] EPA's final PM staff paper and the CASAC review of the EPA staff paper (see references earlier in this report).

[69] For the EPA Administrator's rationale for proposing to retain the current level for the annual $PM_{2.5}$ standard and recognition of the CASAC's recommendation not endorsing this approach, see 71 *Federal Register* 2650-2653, January 17, 2006.

[70] Letter of Dr. Rogene Henderson, Chair, Clean Air Scientific Advisory Committee, to the Hon. Stephen Johnson, Administrator, US EPA, March 21, 2006, EPA-CASAC-LTR-06-002, available at http://yosemite.epa.gov/sab/ sabproduct.nsf/CD706C976DAC62B385257 1390081CC21/$File/casac-ltr-06-002.pdf; or from the federal docket for the proposed rule Docket ID No. EPA-HQ-OAR-2001-0017, on the Federal Docket website http://www. regulations.gov.

[71] CRS communication with Mr. William Becker, Executive Director, State and Territorial Air Pollution Program Administrators/Association of Local Air Pollution Control Officials (STAPPA/ALAPCO), January 5, 2006.

[72] For EPA's discussion and response to several of these comments, see 71 *Federal Register* 61143-61233, October 17, 2006, Part II *Rationale for Final Decisions on Primary $PM_{2.5}$ Standards*, Sections B and F, at http://epa.gov/pm/ actions.html.

[73] CRS communication with Mr. Frank Maisano, Media Contact for the Electric Reliability Coordinating Council, January 17, 2006.

[74] 71 *Federal Register* at 2651, January 17, 2006.

[75] Letter of Dr. Rogene Henderson, Chair of the Clean Air Scientific Advisory Committee et al. to Hon. Stephen L. Johnson, EPA Administrator, September 29, 2006, available at http://yosemite.epa.gov/sab/sabproduct.nsf/ 1C69E987731CB775852571FC00499A10/$File/casac-ltr-06-003.pdf. Italics in original.

[76] Ibid.

[77] Ibid.

[78] See footnote 10.

[79] See footnote 52.

[80] See footnote 46.

[81] See footnote 47.

[82] See Chapter 5 of EPA's October 6, 2006, RIA for more detail, http://www.epa.gov/ ttn/ecas/ria.html.

[83] See Appendix in I *Visibility Benefits Methodology* of EPA's October 6, 2006, RIA http:// www.epa.gov/ttn/ecas/ ria.html.

[84] See footnote 15 and footnote 16.

[85] See footnote 17.

[86] Clean Air Scientific Advisory Committee (CASAC) Review of the EPA Staff Recommendations Concerning a Potential Thoracic Coarse PM Standard in the *Review of the National Ambient Air Quality Standards for Particulate Matter: Policy Assessment of Scientific and Technical Information* (Final PM OAQPS Staff Paper, EPA-452/R-05-005, June 2005), September 15, 2005, EPA-SAB-CASAC-05-007, available on the EPA CASAC website at http://yosemite.epa.gov/sab/sabproduct.nsf/WebReportsbyYearCASAC.

[87] 71 *Federal Register* 61143-61233, October 17, 2006, *Section Vi Reference Methods for the Determination of Particulate Matter as $PM_{10-2.5}$ and $PM_{2.5}$*, http://epa.gov/pm/actions.html.

[88] EPA's standard for PM_{10}, as modified by the 1997 changes to the particulates NAAQS, was challenged shortly after promulgation. Concluding that PM_{10} was a "poorly matched indicator" for thoracic coarse particles because it included the smaller $PM_{2.5}$ category as well as the larger particles, the D.C. Circuit remanded the standard to EPA (American Trucking Associations v. EPA, 175 F.3d 1027, 1054-55 (D.C. Cir. 1999)).

[89] 71 *Federal Register* 61143-61233, October 17, 2006, Section III.C.3. *Decision Not to Revise PM_{10} Indicator*, available at http://epa.gov/pm/actions.html.

[90] See footnote 10.

[91] CASAC reviews, the PM criteria document, staff paper, and related information, are available at http://www.epa.gov/ ttn/naaqs/standards/pm/s_pm_index.html.

[92] Letter of Dr. Rogene Henderson, Chair, Clean Air Scientific Advisory Committee, to the Hon. Stephen Johnson, Administrator, US EPA, March 21, 2006, EPA-CASAC-LTR-06-002, available at http://yosemite.epa.gov/sab/ sabproduct.nsf/CD706C976DAC62B3852571 390081CC21/$File/casac-ltr-06-002.pdf, or from the federal docket for the proposed rule Docket ID No. EPA-HQ-OAR-2001-0017, on the Federal Docket website, http://www.regulations.gov.

[93] Letter to EPA Administrator Stephen L. Johnson, from the Chairman, the Ranking Member, and other Members of the House Committee on Agriculture, July 27, 2006.

[94] 71 *Federal Register* 61143-61233, October 17, 2006, Section III.C.3. *Decision Not to Revise PM_{10} Indicator*, available at http://epa.gov/pm/actions.html.

[95] 71 *Federal Register* 61236-61328, October 17, 2006, at http://epa.gov/pm/actions.html.

[96] Some members of the House Committee on Agriculture expressed their concerns with EPA's final actions with regard to the exemptions at a September 28, 2006, hearing regarding EPA's pesticide programs. House Committee on Agriculture, Subcommittee on Conservation, Credit, Rural Development, and Research, *Review of the Environmental Protection Agency's Pesticide Program*, September 28, 2006.

[97] See footnote 10.

[98] Part IV of the D.C. Circuit's February 24, 2009, decision. The court also did not reach the question raised by the amicus National Association of Home Builders that EPA does not have the authority to set different NAAQS for urban areas or for different emissions sources.

[99] See EPA's PM_{10} designations at http://www.epa.gov/air/oaqps/greenbk/pindex.html.

[100] Information provided directly to CRS by the EPA Office of Air Quality Planning and Standards (OAQPS).

[101] US Environmental Protection Agency, "Release of Final Document Related to the Review of the National Ambient Air Quality Standards for Particulate Matter," 76 *Federal Register* 22665, April 22, 2011, http://www.epa.gov/ttn/naaqs/standards/pm/data/20110419 pmpafinal.pdf.

[102] See assessments referenced in footnote 16.

[103] See footnote 17.

[104] Letters to EPA Administrator Lisa Jackson: joint letter from 21 Senators, July 23, 2010, http://grassley.senate.gov/ about/upload/Agriculture-07-23-10-dust-letter-to-EPA-signed-version-doc.pdf; joint letter from Senators Kent Conrad and Byron Dorgan, and Representative Earl Pomeroy, August 5, 2010, http://conrad.senate.gov/pressroom/ record.cfm? id=327070&.

[105] See footnote 10.

[106] See assessments referenced in footnote 15 and footnote 16. See also US EPA, *Particulate Matter Urban-Focused Visibility Assessment - Second External Review Draft* (EPA 452/P-10-002, January 2010), available at http://www.epa.gov/ttn/naaqs/standards/ pm/s_pm_2007_risk.html.

[107] 71 *Federal Register* 2625, January 17, 2006, http://epa.gov/pm/actions.html.

[108] 71 *Federal Register* 61143-61233, October 17, 2006, http://epa.gov/pm/actions.html.

In: Federal Control of Air Emissions
Editors: H.M. Graham and T.E. Lee
ISBN: 978-1-62081-808-4

Chapter 5

CLEAN AIR ACT: A SUMMARY OF THE ACT AND ITS MAJOR REQUIREMENTS

James E. McCarthy, Claudia Copeland, Larry Parker and Linda-Jo Schierow

SUMMARY

This chapter summarizes the Clean Air Act and its major regulatory requirements.

The principal statute addressing air quality concerns, the Clean Air Act was first enacted in 1955, with major revisions in 1970, 1977, and 1990. The Act requires EPA to set health-based standards for ambient air quality, sets deadlines for the achievement of those standards by state and local governments, and requires EPA to set national emission standards for large or ubiquitous sources of air pollution, including motor vehicles, power plants, and other industrial sources. In addition, the Act mandates emission controls for sources of 187 hazardous air pollutants, establishes a cap-and-trade program to limit acid rain, requires the prevention of significant deterioration of air quality in areas with clean air, requires a program to restore visibility impaired by regional haze in national parks and wilderness areas, and implements the Montreal Protocol to phase out most ozone-depleting chemicals.

This chapter describes the Act's major provisions and provides tables listing all major amendments, with the year of enactment and Public Law number, and cross-referencing sections of the Act with the major U.S. Code sections of the codified statute.

INTRODUCTION

The authorities and responsibilities of the Environmental Protection Agency (EPA) derive primarily from a dozen major environmental statutes. This report provides a concise summary of one of those statutes, the Clean Air Act. It provides a very brief history of federal involvement in air quality regulation and of the provisions added by legislation in 1970, 1977, and 1990; it explains major authorities contained in the Act; it defines key terms; and it lists references for more detailed information on the Act and its implementation.

While this report attempts to present the essence of the Act, it is necessarily incomplete. Many details and secondary provisions are omitted. In addition, the report describes the statute largely without discussing its implementation. Statutory deadlines to control emissions and achieve particular mandates have often been missed as a result of delayed standard-setting by EPA, delayed action on implementation by states and local governments, or law suits brought by interested parties.

OVERVIEW

The Clean Air Act, codified as 42 U.S.C. 7401 *et seq*., seeks to protect human health and the environment from emissions that pollute ambient, or outdoor, air. It requires the Environmental Protection Agency to establish minimum national standards for air quality, and assigns primary responsibility to the states to assure compliance with the standards. Areas not meeting the standards, referred to as "nonattainment areas," are required to implement specified air pollution control measures. The Act establishes federal standards for mobile sources of air pollution and their fuels and for sources of 187 hazardous air pollutants, and it establishes a cap-and-trade program for the emissions that cause acid rain. It establishes a comprehensive permit system for all major sources of air pollution. It also addresses the prevention of pollution in areas with clean air and protection of the stratospheric ozone layer.

Like many other programs administered by the Environmental Protection Agency, federal efforts to control air pollution have gone through several phases, beginning with information collection, research, and technical assistance, before being strengthened to establish federal standards and enforcement. Federal legislation addressing air pollution was first passed in 1955, prior to which air pollution was the exclusive responsibility of state and local levels of government.

The federal role was strengthened in subsequent amendments, notably the Clean Air Act Amendments of 1970, 1977, and 1990. The 1970 amendments established the procedures under which EPA sets national standards for air quality, required a 90% reduction in emissions from new automobiles by 1975, established a program to require the best available control technology at major new sources of air pollution, established a program to regulate air toxics, and greatly strengthened federal enforcement authority. The 1977 amendments adjusted the auto emission standards, extended deadlines for the attainment of air quality standards, and added the Prevention of Significant Deterioration program to protect air cleaner than national standards.

Changes to the Act in 1990 included provisions to (1) classify most nonattainment areas according to the extent to which they exceed the standard, tailoring deadlines, planning, and controls to each area's status; (2) tighten auto and other mobile source emission standards; (3) require reformulated and alternative fuels in the most polluted areas; (4) revise the air toxics section, establishing a new program of technology-based standards and addressing the problem of sudden, catastrophic releases of air toxics; (5) establish an acid rain control program, with a marketable allowance scheme to provide flexibility in implementation; (6) require a state-run permit program for the operation of major sources of air pollutants; (7) implement the Montreal Protocol to phase out most ozone-depleting chemicals; and (8) update the enforcement provisions so that they parallel those in other pollution control acts, including authority for EPA to assess administrative penalties.

The 1990 amendments also authorized appropriations for clean air programs through FY1998. The Act has not been reauthorized since then. House rules require enactment of an authorization before an appropriation bill can be considered; but this requirement can be waived and frequently has been. Thus, while authorization of appropriations in the Clean Air Act (and most other environmental statutes) has expired, programs have continued and have been funded. The Act's other legal authorities, to issue and enforce regulations, are, for the most part, permanent and are not affected by the lack of authorization.

The remainder of this chapter describes major programs required by the Act, with an emphasis on the changes established by the 1990 amendments.

Table 1. Clean Air Act and Amendments (codified generally as 42 U.S.C. 7401-7671)

Year	Act	Public Law Number
1955	Air Pollution Control Act	P.L. 84-159
1959	Reauthorization	P.L. 86-35 3
1960	Motor vehicle exhaust study	P.L. 86-493
1963	Clean Air Act Amendments	P.L. 88-206
1965	Motor Vehicle Air Pollution Control Act	P.L. 89-272, Title I
1966	Clean Air Act Amendments of 1966	P.L. 89-675
1967	Air Quality Act of 1967	P.L. 90-148
	National Air Emission Standards Act	
1970	Clean Air Act Amendments of 1970	P.L. 91-604
1973	Reauthorization	P.L. 93-13
1974	Energy Supply and Environmental Coordination Act of 1974	P.L. 93-319
1977	Clean Air Act Amendments of 1977	P.L. 95-95
1980	Acid Precipitation Act of 1980	P.L. 96-294, Title VII
1981	Steel Industry Compliance Extension Act of 1981	P.L. 97-23
1987	Clean Air Act 8-month Extension	P.L. 100-202
1990	Clean Air Act Amendments of 1990	P.L. 101-549
1991	Technical correction to list of hazardous air pollutants	P.L. 102-187
1995-96	Relatively minor laws amending the Act	P.L. 104-6, 59, 70, 260
1998	Amended § 604 re methyl bromide	P.L. 105-277, Section 764
1998	Border Smog Reduction Act of 1998	P.L. 105-286
1999	Chemical Safety Information, Site Security and Fuels Regulatory Relief Act	P.L. 106-40
2004	Amendments to §209 re small engines	P.L. 108-199, Division G, Title IV, Section 428
2005	Energy Policy Act of 2005 (amended §211 re fuels)	P.L. 109-58
2007	Energy Independence and Security Act of 2007 (amended §211 re fuels)	P.L. 110-140

NATIONAL AMBIENT AIR QUALITY STANDARDS

In section 109, the Act requires EPA to establish National Ambient Air Quality Standards (NAAQS) for air pollutants that endanger public health or welfare, in the Administrator's judgment, and whose presence in ambient air results from numerous or diverse sources. The NAAQS must be designed to protect public health and welfare with an adequate margin of safety. Using this authority, EPA has promulgated NAAQS for six air pollutants: sulfur dioxide (SO_2), particulate matter (PM2.5 and PM10), nitrogen dioxide (NO_2), carbon monoxide (CO), ozone,[1] and lead. The Act requires EPA to review the scientific data upon which the standards are based, and revise the standards, if necessary, every five years. More often than not, however, EPA has taken more than five years in reviewing and revising the standards.

Originally, the Act required that the NAAQS be attained by 1977 at the latest, but the states experienced widespread difficulty in complying with this deadline. As a result, the deadlines for achieving NAAQS have been extended several times. Under the 1990 amendments, most areas not in attainment with NAAQS must meet special compliance schedules, staggered according to the severity of an area's air pollution problem. The amendments also established specific requirements for each nonattainment category, as described below.

STATE IMPLEMENTATION PLANS

While the Act authorizes the EPA to set NAAQS, the states are responsible for establishing procedures to attain and maintain the standards. Under Section 110 of the Act, the states adopt plans, known as State Implementation Plans (SIPs), and submit them to EPA to ensure that they are adequate to meet statutory requirements.

SIPs are based on emission inventories and computer models to determine whether air quality violations will occur. If these data show that standards would be exceeded, the state must impose additional controls on existing sources to ensure that emissions do not cause "exceedances" of the standards. Proposed new and modified sources must obtain state construction permits in which the applicant shows how the anticipated emissions will not exceed allowable limits. In nonattainment areas, emissions from new or modified sources must also be offset by reductions in emissions from existing sources.

The 1990 amendments require EPA to impose sanctions in areas which fail to submit a SIP, fail to submit an adequate SIP, or fail to implement a SIP: unless the state corrects such failures, a 2-to-1 emissions offset for the construction of new polluting sources is imposed 18 months after notification to the state, and a ban on most new federal highway grants is imposed six months later. An additional ban on air quality grants is discretionary. Ultimately, a Federal Implementation Plan may be imposed if the state fails to submit or implement an adequate SIP.

The amendments also require that, in nonattainment areas, no federal permits or financial assistance may be granted for activities that do not "conform" to a State Implementation Plan. This requirement can cause a temporary suspension in funding for most new highway and transit projects if an area fails to demonstrate that the emissions caused by such projects are consistent with attainment and maintenance of ambient air quality standards. Demonstrating conformity of transportation plans and SIPs is required in nonattainment areas whenever new plans are submitted.

Nonattainment Requirements

In a major departure from the prior law, the 1990 Clean Air Act Amendments grouped most nonattainment areas into classifications based on the extent to which the NAAQS was exceeded, and established specific pollution controls and attainment dates for each classification. These requirements are described here as spelled out in Sections 181-193 of the Act.[2]

Nonattainment areas are classified on the basis of a "design value," which is derived from the pollutant concentration (in parts per million or micrograms per cubic meter) recorded by air quality monitoring devices. The design value for the 1-hour ozone standard was the fourth highest hourly reading measured during the most recent three-year period. Using these design values, the Act created five classes of ozone nonattainment, as shown in Table 2. Only Los Angeles fell into the "extreme" class, but 97 other areas were classified in one of the other four ozone categories. A simpler classification system established moderate and serious nonattainment areas for carbon monoxide and particulate matter with correspondingly more stringent control requirements for the more polluted class.

As shown in the table, the deadlines for attainment for ozone nonattainment areas stretched from 1993 to 2010, depending on the severity of the problem. (Under the 8-hour ozone standard, which replaced the 1-hour standard in 2004, these deadlines are changed to 2007 to 2021.) For carbon monoxide, the attainment date for moderate areas was December 31, 1995, and for serious areas, December 31, 2000. For particulate matter, the deadline for areas designated moderate nonattainment as of 1990 was December 31, 1994; for those areas subsequently designated as moderate, the deadline is six years after designation. For serious areas, the respective deadlines are December 31, 2001 or 10 years after designation.

Table 2. Ozone Nonattainment Classifications

Class	Marginal	Moderate	Serious	Severe	Extreme
Deadline	1993	1996	1999	2005-2007[a]	2010
Areas[b]	42 areas	32 areas	14 areas	9 areas	1 area
Design Value	0.121 ppm-0.138 ppm	0.138 ppm-0.160 ppm	0.160 ppm-0.180 ppm	0.180 ppm-0.280 ppm	>0.280 ppm

a. Areas with a 1988 design value between 0.190 and 0.280 ppm have 17 years to attain; others have 15 years.

b. Number of areas in each category as of the date of enactment.

Requirements for Ozone Nonattainment Areas

Although areas with more severe air pollution problems have a longer time to meet the standards, more stringent control requirements are imposed in areas with worse pollution. A summary of the primary ozone control requirements for each nonattainment category follows.

Marginal Areas

- Inventory emissions sources (to be updated every three years).
- Require 1.1 to 1 offsets (i.e., new major emission sources of volatile organic compounds [VOCs] must reduce VOC emissions from existing facilities in the area by 10% more than the emissions of the new facility).
- Impose reasonably available control technology (RACT) on all major sources emitting more than 100 tons per year for the nine industrial categories where EPA had already issued control technique guidelines describing RACT prior to 1990.

Moderate Areas

- Meet all requirements for marginal areas.
- Impose a 15% reduction in VOC emissions in six years.
- Adopt a basic vehicle inspection and maintenance program.
- Impose RACT on all major sources emitting more than 100 tons per year for all additional industrial categories where EPA will issue control technique guidelines describing RACT.
- Require vapor recovery at gas stations selling more than 10,000 gallons per month.
- Require 1.15 to 1 offsets.

Serious Areas

- Meet all requirements for moderate areas.
- Reduce definition of a major source of VOCs from emissions of 100 tons per year to 50 tons per year for the purpose of imposing RACT.
- Reduce VOCs 3% annually for years 7 to 9 after the 15% reduction already required by year 6.
- Improve monitoring.
- Adopt an enhanced vehicle inspection and maintenance program.
- Require fleet vehicles to use clean alternative fuels.
- Adopt transportation control measures if the number of vehicle miles traveled in the area is greater than expected.
- Require 1.2 to 1 offsets.
- Adopt contingency measures if the area does not meet required VOC reductions.

Severe Areas

- Meet all requirements for serious areas.
- Reduce definition of a major source of VOCs from emissions of 50 tons per year to 25 tons per year for the purpose of imposing RACT.
- Adopt specified transportation control measures.
- Implement a reformulated gasoline program.
- Require 1.3 to 1 offsets.
- Impose $5,000 per ton penalties on major sources if the area does not meet required reductions.

Extreme Areas

- Meet all requirements for severe areas.
- Reduce definition of a major source of VOCs from emissions of 25 tons per year to 10 tons per year for the purpose of imposing RACT.
- Require clean fuels or advanced control technology for boilers emitting more than 25 tons per year of NO_x.
- Require 1.5 to 1 offsets.

As noted, EPA promulgated a new, 8-hour ozone standard in July 1997. Following extensive court challenges, the agency designated nonattainment areas for the new standard on April 30, 2004. State Implementation Plans were required to be submitted in 2007.

Requirements for Carbon Monoxide Nonattainment Areas

As with ozone nonattainment areas, carbon monoxide (CO) nonattainment areas are subjected to specified control requirements, with more stringent requirements in Serious nonattainment areas. A summary of the primary CO control requirements for each nonattainment category follows.

Moderate Areas

- Conduct an inventory of emissions sources.
- Forecast total vehicle miles traveled in the area.
- Adopt an enhanced vehicle inspection and maintenance program.
- Demonstrate annual improvements sufficient to attain the standard.

Serious Areas

- Adopt specified transportation control measures.
- Implement an oxygenated fuels program for all vehicles in the area.
- Reduce definition of a major source of CO from emissions of 100 tons per year to 50 tons per year if stationary sources contribute significantly to the CO problem.

Serious areas failing to attain the standard by the deadline have to revise their SIP and demonstrate reductions of 5% per year until the standard is attained.

Requirements for Particulate Nonattainment Areas

Particulate (PM10) nonattainment areas are also subject to specified control requirements. These are:

Moderate Areas

- Require permits for new and modified major stationary sources of PM10.
- Impose reasonably available control measures (RACM).

Serious Areas

- Impose best available control measures (BACM).
- Reduce definition of a major source of $_{\text{PM10}}$ from 100 tons per year to 70 tons per year.

In July 1997, EPA promulgated new standards for fine particulates $_{\text{(PM2.5)}}$. The $_{\text{PM2.5}}$ standards were also subject to court challenges. The absence of a monitoring network capable of measuring the pollutant delayed implementation as well. Nonattainment areas for $_{\text{PM2.5}}$ were designated on April 14, 2005. States had three years subsequent to designation to submit State Implementation Plans. Revisions to the NAAQS promulgated in October 2006 strengthened the $_{\text{PM2.5}}$ standard.

TRANSPORTED AIR POLLUTION

Meeting the nation's clean air standards can be complicated as air pollution is no respecter of political boundaries or subdivisions. This problem of transported air pollutants has come into particular focus as states and EPA attempt to develop effective compliance strategies to achieve both the ozone and the PM2.5 NAAQS. Under Section 110(a)(2)(D), SIPs must include adequate provisions to prevent sources within that state from contributing significantly to nonattainment in one or more downwind states.

If EPA finds a SIP inadequate to achieve a NAAQS, it must require the affected state to submit a revised SIP that includes sufficient measures to bring that state into compliance. This is known as a "SIP Call." The 1990 Clean Air Act amendments provided EPA and the states with new tools to address the transport problem through this provision. One of those tools is Section 176A, a provision that permits the EPA, either on its own or by petition from any state, to establish a transport region to address regional pollution problems contributing to violations of a primary NAAQS. A commission of EPA and state officials is constituted to make recommendations to EPA on appropriate mitigation strategies. Based on the commission's findings and recommendations, EPA is then required under section 1 10(k)(5) to notify affected states of inadequacies in their current state implementation plans and to establish deadlines (not to exceed 18 months) for submitting necessary revisions (i.e., a SIP call). Besides authorizing administratively-created transport regions, the 1990 amendments statutorily created an Ozone Transport Region (OTR) in the Northeast. This provision (Section 184 of the act) required specific additional controls for all areas (not only nonattainment areas) in that region, and established the Ozone Transport Commission for the purpose of recommending to EPA regionwide controls affecting all areas in the region.

The transport issue may also be addressed by affected downwind states through a Section 126 petition. As amended by the 1990 Clean Air Act amendments, under Section 126(b) any state or political subdivision may petition EPA for a finding that a major source or group of stationary sources located in another state is emitting pollutants that "significantly contribute" to the nonattainment of a NAAQS by their state. EPA is to respond to the petition within 60 days. If the petition is granted, the offending sources must cease operations within three months unless the sources comply with emission controls and the compliance schedules as determined by EPA to bring them into compliance with the section. Section 126 has rarely been used, although it has proven useful to EPA in some cases as backup authority where there might be challenges to a SIP call.

EMISSION STANDARDS FOR MOBILE SOURCES

Title II of the Clean Air Act has required emission standards for automobiles since 1968. The 1990 amendments significantly tightened these standards: for cars, the hydrocarbon standard was reduced by 40% and the nitrogen oxides (NO_x) standard by 50%. These standards—referred to as "Tier 1" standards—were phased in over the 1994-1996 model years.

The amendments envisioned a further set of reductions ("Tier 2" standards), but not before model year 2004. For Tier 2 standards to be promulgated, the agency was first required to report to Congress concerning the need for further emission reductions, the availability of technology to achieve such reductions, and the cost-effectiveness of such controls compared to other means of attaining air quality standards. EPA submitted this report to Congress in August 1998, concluding that further emission reductions were needed and that technology to achieve such reductions was available and cost-effective. Tier 2 standards, requiring emission reductions of 77% to 95% from cars and light trucks were promulgated in February 2000, and were phased in over the 2004-2009 model years. To facilitate the use of more effective emission controls, the standards also require a more than 90% reduction in the sulfur content of gasoline, beginning in 2004.

The 1990 amendments also required that oxygenated gasoline, designed to reduce emissions of carbon monoxide, be sold in the worst CO nonattainment areas and that "reformulated" gasoline (RFG), designed to reduce emissions of volatile organic compounds and toxic air pollutants, be sold in the nine worst ozone nonattainment areas (Los Angeles, San Diego, Houston, Baltimore, Philadelphia, New York, Hartford, Chicago, and Milwaukee); metropolitan Washington, DC; and four areas in California were added to the mandatory list later. Other ozone nonattainment areas can opt in to the RFG program; as of 2006, additional areas in 11 states had done so.

The fuels provisions were modified by the Energy Policy Act of 2005 (EPACT), removing the requirement that RFG contain oxygenates. Instead, EPACT required the use of increasing amounts of renewable fuel, most likely to be ethanol, in motor fuels, beginning in 2006. The Energy Independence and Security Act of 2007 further strengthened the renewable fuel requirements.

Use of alternative fuels and development of cleaner engines was also to be stimulated by the Clean-Fuel Fleet Program. In all of the most seriously polluted ozone and CO nonattainment areas, centrally fueled fleets of 10 or more passenger cars and light-duty trucks must purchase at least 30% clean-fuel vehicles when they add new vehicles to existing fleets, starting in 1999. (The Act originally required the program to begin in 1998, but the start was delayed by a year.) The percentage rose to 50% in 2000 and 70% in 2001. Heavy-duty fleets are required to purchase at least 50% clean-fuel vehicles annually. A clean fuel vehicle is one which meets Low Emission Vehicle (LEV) standards and operates on reformulated gasoline, reformulated diesel, methanol, ethanol, natural gas, liquefied petroleum gas, hydrogen, or electricity.

In addition to the above program, California's Zero Emission Vehicle (ZEV) program also is intended to promote the development of alternative fuels and vehicles. Section 209(b) of the Clean Air Act allows the EPA Administrator to grant California the authority to develop its own vehicle emissions standards if those standards are at least as stringent as the federal standards and if the state demonstrates that it needs the standards to meet compelling and extraordinary conditions. In addition to setting more stringent standards for all vehicles, California used this authority to establish a program requiring auto manufacturers to sell ZEVs (electric or hydrogen fuel cell vehicles) in the state beginning in 2003. This program has been substantially modified since it was enacted, and now allows credit for hybrid and partial ZEV vehicles in addition to true ZEVs, but it has served as an incubator for lower emission technologies since its adoption. Section 177 of the Act allows other states to adopt California's stricter standards: at least ten states (Connecticut, Maine, Massachusetts, New Jersey, New York, Oregon, Pennsylvania, Rhode Island, Vermont, and Washington) have already adopted them or are in the process of doing so.

The 1990 amendments also imposed tighter requirements on certification (an auto's useful life is defined as 100,000 miles instead of the earlier 50,000 miles), on emissions allowed during refueling, on low temperature CO emissions, on in-use performance over time, and on warranties for the most expensive emission control components (8 years/80,000 miles for the catalytic converter, electronic emissions control unit, and onboard emissions diagnostic unit). Regulations were also extended to include nonroad fuels and engines.

Standards for trucks and buses using diesel engines were also strengthened. The 1990 amendments required new urban buses to reduce emissions of diesel particulates 92% by 1996, and all other heavy-duty diesel engines to achieve an 83% reduction by the same year. NO_x emissions must also be reduced, 33% by 1998. Authority to further strengthen these standards led to promulgation in January 2001 of new emission standards requiring a further 90%-95% reduction in emissions phased in over the 2007-2010 model years, and a reduction of 97% in the allowable amount of sulfur in highway diesel fuel. These regulations were followed in May 2004 by similar requirements for nonroad diesel equipment, which will be phased in between 2007 and 2015.

Hazardous Air Pollutants[3]

Completely rewritten by the Clean Air Act Amendments of 1990, Section 112 of the Act establishes programs for protecting public health and the environment from exposure to toxic air pollutants. As revised by the 1990 amendments, the section contains four major provisions: Maximum Achievable Control Technology (MACT) requirements; health-based standards; standards for stationary "area sources" (small, but numerous sources, such as gas stations or dry cleaners, that collectively emit significant quantities of hazardous pollutants); and requirements for the prevention of catastrophic releases.

First, EPA is to establish technology-based emission standards, called MACT standards, for sources of 187 pollutants listed in the legislation, and to specify categories of sources subject to the emission standards.[4] EPA is to revise the standards periodically (at least every eight years). EPA can, on its own initiative or in response to a petition, add or delete substances or source categories from the lists.

Section 112 establishes a presumption in favor of regulation for the designated chemicals; it requires regulation of a designated pollutant unless EPA or a petitioner is able to show "that there is adequate data on the health and environmental effects of the substance to determine that emissions, ambient concentrations, bioaccumulation or deposition of the substance may not reasonably be anticipated to cause any adverse effects to human health or adverse environmental effects."

EPA is required to set standards for sources of the listed pollutants that achieve "the maximum degree of reduction in emissions" taking into account cost and other non-air-quality factors. These MACT standards for new sources "shall not be less stringent than the most stringent emissions level that is achieved in practice by the best controlled similar source." The standards for existing sources may be less stringent than those for new sources, but must be no less stringent than the emission limitations achieved by either the best performing 12% of existing sources (if there are more than 30 such sources in the category or subcategory) or the best performing 5 similar sources (if there are fewer than 30). Existing sources are given three years following promulgation of standards to achieve compliance, with a possible one-year extension; additional extensions may be available for special circumstances or for certain categories of sources. Existing sources that achieve voluntary early emissions reductions will receive a six-year extension for compliance with MACT.

The second major provision of Section 112 directs EPA to set health-based standards to address situations in which a significant residual risk of adverse health effects or a threat of adverse environmental effects remains after installation of MACT. This provision requires that EPA, after consultation with the Surgeon General of the United States, submit a report to Congress on the public health significance of residual risks, and recommend legislation regarding such risks. If Congress does not legislate in response to EPA's recommendations, then EPA is required to issue standards for categories of sources of hazardous air pollutants as necessary to protect the public health with an ample margin of safety or to prevent an adverse environmental effect. A residual risk standard is required for any source emitting a cancer-causing pollutant that poses an added risk to the most exposed person of more than one-in-a-million. Residual risk standards are due eight years after promulgation of MACT for the affected source category. Existing sources have 90 days to comply with a residual risk standard, with a possible two-year extension. In general, residual risk standards do not apply to area sources.

The law directed EPA to contract with the National Academy of Sciences (NAS) for a study of risk assessment methodology, and created a Risk Assessment and Management Commission to investigate and report on policy implications and appropriate uses of risk assessment and risk management. In 1994 NAS published its report, *Science and Judgment in Risk Assessment.* The Commission study, *Framework for Environmental Health Risk Management*, was released in 1997.

Third, in addition to the technology-based and health-based programs for major sources of hazardous air pollution, EPA is to establish standards for stationary "area sources" determined to present a threat of adverse effects to human health or the environment. The provision requires EPA to regulate the stationary area sources responsible for 90% of the emissions of the 30 hazardous air pollutants that present the greatest risk to public health in the largest number of urban areas. In setting the standard, EPA can impose less stringent "generally available" control technologies, rather than MACT.

Finally, Section 112 addresses prevention of sudden, catastrophic releases of air toxics by establishing an independent Chemical Safety and Hazard Investigation Board. The Board is responsible for investigating accidents involving releases of hazardous substances, conducting studies, and preparing reports on the handling of toxic materials and measures to reduce the risk of accidents.

EPA is also directed to issue prevention, detection, and correction requirements for catastrophic releases of air toxics by major sources. Section 112(r) requires owners and operators to prepare risk management plans including hazard assessments, measures to prevent releases, and a response program.

New Source Performance Standards[5]

Section 111 of the Act requires EPA to establish nationally uniform, technology-based standards (called New Source Performance Standards, or NSPS) for categories of new industrial facilities. These standards accomplish two goals: first, they establish a consistent baseline for pollution control that competing firms must meet, and thereby remove any incentive for states or communities to weaken air pollution standards in order to attract polluting industry; and second, they preserve clean air to accommodate future growth, as well as for its own benefits.

NSPS establish maximum emission levels for new major stationary sources—powerplants, steel mills, and smelters, for example—with the emission levels determined by the best "adequately demonstrated" continuous control technology available, taking costs into account. EPA must regularly revise and update NSPS applicable to designated sources as new technology becomes available, since the goal is to prevent new pollution problems from developing and to force the installation of new control technology.

The standards also apply to modifications of existing facilities, through a process called New Source Review (NSR). The law's ambiguity regarding what constitutes a modification (subject to NSR) as opposed to routine maintenance of a facility has led to litigation, with EPA proposing in recent years to modify its interpretation of the requirements of this section.

Solid Waste Incinerators

Prior to 1990, solid waste incinerators, which emit a wide range of pollutants, were subject to varying degrees of state and federal regulation depending on their size, age, and the type of waste burned. In a new Section 129, the 1990 amendments established more consistent federal requirements specifying that emissions of 10 categories of pollutants be regulated at new and existing incinerators burning municipal solid waste, medical waste, and commercial and industrial waste. The amendments also established emissions monitoring and operator training requirements.

Prevention of Significant Deterioration/Regional Haze[6]

Sections 160-169 of the act establish requirements for the prevention of significant deterioration of air quality (PSD). The PSD program reflects the principle that areas where air quality is better than that required by NAAQS should be protected from significant new air pollution even if NAAQS would not be violated.

The Act divides clean air areas into three classes, and specifies the increments of SO_2 and particulate pollution allowed in each. Class I areas include international and national parks, wilderness and other pristine areas; allowable increments of new pollution are very small. Class II areas include all attainment and not classifiable areas, not designated as Class I; allowable increments of new pollution are modest. Class III represents selected areas that states may designate for development; allowable increments of new pollution are large (but not exceeding NAAQS). Through an elaborate hearing and review process, a state can have regions redesignated from Class II to Class III (although none have yet been so redesignated).

While the 1977 amendments only stipulated PSD standards for two pollutants, SO_2 and particulates, EPA is supposed to establish standards for other criteria pollutants. Thus far, only one of the other four (NO_2) has been addressed: the agency promulgated standards for NO_2 in 1988.

Newly constructed polluting sources in PSD areas must install best available control technology (BACT) that may be more strict than that required by NSPS. The justifications of the policy are that it protects air quality, provides an added margin of health protection, preserves clean air for future development, and prevents firms from gaining a competitive edge by "shopping" for clean air to pollute.

In Sections 1 69A and B, the Act also sets a national goal of preventing and remedying impairment of visibility in national parks and wilderness areas, and requires EPA to promulgate regulations to assure reasonable progress toward that goal. In the 1990 Amendments, Congress strengthened these provisions, which had not been implemented.

The amendments required EPA to establish a Grand Canyon Visibility Transport Commission, composed of Governors from each state in the affected region, an EPA designee, and a representative of each of the national parks or wilderness areas in the region. Other visibility transport commissions can be established upon EPA's discretion or upon petition from at least two states.

Within 18 months of receiving a report from one of these commissions, EPA is required to promulgate regulations to assure reasonable progress toward the visibility goal, including requirements that states update their State Implementation Plans to contain emission limits, schedules of compliance, and other measures necessary to make reasonable progress. Specifically mentioned is a requirement that states impose Best Available Retrofit Technology on existing sources of emissions impairing visibility.

The Grand Canyon Commission delivered a set of recommendations to EPA in June 1996, and the agency subsequently promulgated a “regional haze” program applicable to all 50 states under this authority.

ACID DEPOSITION CONTROL[7]

The Clean Air Act Amendments of 1990 added an acid deposition control program (Title IV) to the Act. It set goals for the year 2000 of reducing annual SO_2 emissions by 10 million tons from 1980 levels and reducing annual NO_x emissions by 2 million tons, also from 1980 levels.

The SO_2 reductions were imposed in two steps. Under Phase 1, owners/operators of 111 electric generating facilities listed in the law that arc larger than 100 megawatts had to meet tonnage emission limitations by January 1, 1995. This would reduce SO_2 emission by about 3.5 million tons. Phase 2 included facilities larger than 75 megawatts, with a deadline of January 1, 2000. Compliance has been 100%.

To introduce some flexibility in the distribution and timing of reductions, the Act creates a comprehensive permit and emissions allowance system. An allowance is a limited authorization to emit a ton of SO_2. Issued by EPA, the allowances would be allocated to Phase 1 and Phase 2 units in accordance with baseline emissions estimates. Powerplants which commence operation after November 15, 1990 would not receive any allowances. These new units would have to obtain allowances (offsets) from holders of existing allowances. Allowances may be traded nationally during either phase. The law also permitted industrial sources and powerplants to sell allowances to utility systems under regulations developed by EPA. Allowances may be banked by a utility for future use or sale.

The Act provided for two types of sales to improve the liquidity of the allowance system and to ensure the availability of allowances for utilities and independent power producers who need them. First, a special reserve fund consisting of 2.8% of Phase 1 and Phase 2 allowance allocations has been set aside for sale. Allowances from this fund (25,000 annually from 1993-1999 and 50,000 thereafter) are sold at a fixed price of $1,500 an allowance. Independent power producers have guaranteed rights to these allowances under certain conditions. Second, an annual, open auction sold allowances (150,000 from 1993-1995, and 250,000 from 1996-1999) with no minimum price. Utilities with excess allowances may have them auctioned off at this auction, and any person may buy allowances.

The Act essentially caps SO_2 emissions at individual existing sources through a tonnage limitation, and at future plants through the allowance system. First, emissions from most existing sources are capped at a specified emission rate times an historic baseline level. Second, for plants commencing operation after November 15, 1990, emissions must be completely offset with additional reductions at existing facilities beginning after Phase 2 compliance. However, as noted above, the law provides some allowances to future powerplants which meet certain criteria. The utility SO_2 emission cap was set at 8.9 million tons, with some exceptions.

The Act provides that if an affected unit does not have sufficient allowances to cover its emissions, it is subject to an excess emission penalty of $2,000 per ton of SO_2 and required to reduce an additional ton of SO_2 the next year for each ton of excess pollutant emitted.

The Act also requires EPA to inventory industrial emissions of SO_2 and to report every five years, beginning in 1995. If the inventory shows that industrial emissions may reach levels above 5.60 million tons per year, then EPA is to take action under the Act to ensure that the 5.60 million ton cap is not exceeded. The Act requires EPA to set specific NO_x emission rate limitations—0.45 lb. per million Btu for tangentially-fired boilers and 0.50 lb. per million Btu for wall-fired boilers—unless those rates can not be achieved by low-NO_x burner technology. Tangentially and wall-fired boilers affected by Phase 1 SO_2 controls must also meet NO_x requirements. EPA was to set emission limitations for other types of boilers by 1997 based on low-NO_x burner costs, which EPA did. In addition, EPA was to propose and promulgate a revised new source performance standard for NO_x from fossil fuel steam generating units, which EPA also did, in 1998.

PERMITS[8]

The Clean Air Act Amendments of 1990 added a Title V to the Act which requires states to administer a comprehensive permit program for the operation of sources emitting air pollutants. These requirements are modeled after similar provisions in the Clean Water Act. Previously, the Clean Air Act contained limited provision for permits, requiring only new or modified major stationary sources to obtain construction permits (under Section 165 of the Act).

Sources subject to the permit requirements generally include major sources that emit or have the potential to emit 100 tons per year of any regulated pollutant, plus stationary and area sources that emit or have potential to emit lesser specified amounts of hazardous air pollutants. However, in nonattainment areas, the permit requirements also include sources which emit as little as 50, 25, or 10 tons per year of VOCs, depending on the severity of the region's nonattainment status (serious, severe, or extreme).

States were required to develop permit programs and to submit those programs for EPA approval by November 15, 1993. EPA had one year to approve or disapprove a state's submission in whole or in part. After the effective date of a state plan, sources had 12 months to submit an actual permit application.

States are to collect annual fees from sources sufficient to cover the "reasonable costs" of administering the permit program, with revenues to be used to support the agency's air pollution control program. The fee must be at least $25 per ton of regulated pollutants (excluding carbon monoxide). Permitting authorities have discretion not to collect fees on emissions in excess of 4,000 tons per year and may collect other fee amounts, if appropriate.

The permit states how much of which air pollutants a source is allowed to emit. As a part of the permit process, a source must prepare a compliance plan and certify compliance. The term of permits is limited to no more than five years; sources are required to renew permits at that time. State permit authorities must notify contiguous states of permit applications that may affect them; the application and any comments of contiguous states must be forwarded to EPA for review. EPA can veto a permit; however, this authority is essentially limited to major permit changes. EPA review need not include permits which simply codify elements of a state's overall clean air plan, and EPA has discretion to not review permits for small sources. Holding a permit to some extent shields a source from enforcement actions: the Act provides that a source cannot be held in violation if it is complying with explicit requirements addressed in a permit, or if the state finds that certain provisions do not apply to that source.

ENFORCEMENT

Section 113 of the Act, which was also strengthened by the 1990 amendments, covers enforcement. The section establishes federal authority to issue agency and court orders requiring compliance and to impose penalties for violations of Act requirements. Section 114 authorizes EPA to require sources to submit reports, monitor emissions, and certify compliance with the Act's requirements, and authorizes EPA personnel to conduct inspections.

Like most federal environmental statutes, the Clean Air Act is enforced primarily by states or local governments; they issue most permits, monitor compliance, and conduct the majority of inspections. The federal government functions as a backstop, with authority to review state actions. The agency may act independently or may file its own enforcement action in cases where it concludes that a state's response was inadequate.

The Act also provides for citizen suits both against persons (including corporations or government agencies) alleged to have violated emissions standards or permit requirements, and against EPA in cases where the Administrator has failed to perform an action that is not discretionary under the Act. Citizen groups have often used the latter provision to compel the Administrator to promulgate regulations required by the statute.

The 1990 Amendments elevated penalties for some knowing violations from misdemeanors to felonies; removed the ability of a source to avoid an enforcement order or civil penalty by ceasing a violation within 60 days of notice; gave authority to EPA to assess administrative penalties; and authorized $10,000 awards to persons supplying information leading to convictions under the Act.

STRATOSPHERIC OZONE PROTECTION[9]

Title VI of the 1990 Clean Air Act Amendments represents the United States' primary response on the domestic front to the ozone depletion issue. It also implements the U.S. international responsibilities under the Montreal Protocol on Substances that Deplete the Ozone Layer (and its amendments). Indeed, Section 606(a)(3) provides that the Environmental Protection Agency shall adjust phase-out schedules for ozone-depleting substances in accordance with any future changes in Montreal Protocol schedules. As a result, the phase-out schedules contained in Title VI for various ozone-depleting compounds have now been superseded by subsequent amendments to the Montreal Protocol. Since passage of Title VI, depleting substances such as CFCs, methyl chloroform, carbon tetrachloride, and halons (referred to as Class 1 substances) have been phased out by industrial countries, including the United States. New uses of hydrochlorofluorocarbons (HCFCs) (called Class 2 substances under Title VI) are banned beginning January 1, 2015, unless the HCFCs are recycled, used as a feedstock, or used as a refrigerant for appliances manufactured prior to January 1, 2020. Production of HCFCs is to be frozen January 1, 2015 and phased out by January 1, 2030. Exemptions consistent with the Montreal Protocol are allowed.

The EPA is required to add any substance with an ozone depletion potential (ODP) of 0.2 or greater to the list of Class 1 substances and set a phase-out schedule of no more than seven years. For example, methyl bromide (ODP estimated by EPA at 0.7) was added to the list in December 1993, requiring its phaseout by January 1, 2001; this decision was altered by Congress in 1998 to harmonize the U.S. methyl bromide phase-out schedule with the 2005 deadline set by the parties to the Montreal Protocol in 1997. Also, EPA is required to add any substance that is known or may be reasonably anticipated to harm the stratosphere to the list of Class 2 substances and set a phase-out schedule of no more than ten years.

Title VI contains several implementing strategies to avoid releases of ozone-depleting chemicals to the atmosphere, including (1) for Class 1 substances used as refrigerant—lowest achievable level of use and emissions, maximum recycling, and safe disposal required by July 1, 1992; (2) for servicing or disposing refrigeration equipment containing Class 1 and 2 substances—venting banned as of July 1, 1992; (3) for motor vehicle air conditioners containing Class 1 or 2 substances—recycling required by January 1, 1992 (smaller shops by January 1, 1993); (4) sale of small containers of class 1 and 2 substances—banned within two years of enactment; and (5) nonessential products—banned within two years of enactment.

Selected References

[1] U.S. Environmental Protection Agency. Office of Air Quality Planning and Standards. *Air Trends*. Research Triangle Park, NC. Compiled annually, and available at http://www.epa.gov/airtrends/.

[2] Martineau, Robert J., Jr. & Novello, David P. (eds.) (2004). *The Clean Air Act Handbook*. 2nd edition. Chicago: American Bar Association. 728 p.

Table 3. Major U.S. Code Sections of the Clean Air Act (Codified Generally as 42 U.S.C. 7401-7671)

42 U.S.C.	Section Title	Clean Air Act, as amended
Subchapter I -	Programs and Activities	
Part A -	Air Quality Emissions and Limitations	
7401	Findings, purpose	Sec. 101
7402	Cooperative activities	Sec. 102
7403	Research, investigation, training	Sec. 103
7404	Research relating to fuels and vehicles	Sec. 104
7405	Grants for air pollution planning and control programs	Sec. 105
7406	Interstate air quality agencies; program cost limitations	Sec. 106
7407	Air quality control regions	Sec. 107
7408	Air quality criteria and control techniques	Sec. 108
7409	National primary and secondary air quality standards	Sec. 109
7410	SIPs for national primary and secondary air quality standards	Sec. 110
7411	Standards of performance for new stationary sources	Sec. 111
7412	Hazardous air pollutants	Sec. 112
7413	Federal enforcement	Sec. 113
7414	Recordkeeping, inspections, monitoring, and entry	Sec. 114
7415	International air pollution	Sec. 115
7416	Retention of state authority	Sec. 116
7417	Advisory committees	Sec. 117
7418	Control of pollution from federal facilities	Sec. 118
7419	Primary nonferrous smelter orders	Sec. 119
7420	Noncompliance penalty	Sec. 120
7421	Consultation	Sec. 121
7422	Listing of certain unregulated pollutants	Sec. 122
7423	Stack heights	Sec. 123
7424	Assurance of adequacy of state plans	Sec. 124
7425	Measures to prevent economic disruption/unemployment	Sec. 125

Table 3. (Continued)

42 U.S.C.	Section Title	Clean Air Act, as amended
7426	Interstate pollution abatement	Sec. 126
7427	Public notification	Sec. 127
7428	State boards	Sec. 128
7429	Solid waste combustion	Sec. 129
7430	Emission factors	Sec. 130
7431	Land use authority	Sec. 131
Part B - Ozone Protection (repealed—new provisions related to stratospheric ozone protection are found at 42 U.S.C. 7671 et seq., below)		
Part C - Prevention of Significant Deterioration of Air Quality		
Subpart I -	Clean Air	
7470	Congressional declaration of purpose	Sec. 160
7471	Plan requirements	Sec. 161
7472	Initial classifications	Sec. 162
7473	Increments and ceilings	Sec. 163
7474	Area redesignation	Sec. 164
7475	Preconstruction requirements	Sec. 165
7476	Other pollutants	Sec. 166
7477	Enforcement	Sec. 167
7478	Period before plan approval	Sec. 168
7479	Definitions	Sec. 169
Subpart II -	Visibility Protection	
7491	Visibility protection for federal class I areas	Sec. 169A
7492	Visibility	Sec. 169B
Part D -	Plan Requirements for Nonattainment Areas	
Subpart 1 -	Nonattainment Areas in General	
7501	Definitions	Sec. 171
7502	Nonattainment plan provisions in general	Sec. 172
7503	Permit requirements	Sec. 173
7504	Planning procedures	Sec. 174
7505	Environmental Protection Agency grants	Sec. 175
7505a	Maintenance plans	Sec. 175A
7506	Limitations on certain federal assistance	Sec. 176
7506a	Interstate transport commissions	Sec. 176A
7507	New motor vehicle emission standards in nonattainment areas	Sec. 177
7508	Guidance documents	Sec. 178
7509	Sanctions and consequences of failure to attain	Sec. 179
7509a	International border areas	Sec. 179B

42 U.S.C.	Section Title	Clean Air Act, as amended
Subpart 2 -	Additional Provisions for Ozone Nonattainment Areas	
7511	Classifications and attainment dates	Sec. 181
7511a	Plan submissions and requirements	Sec. 182
7511b	Federal ozone measures	Sec. 183
7511c	Control of interstate ozone air pollution	Sec. 184
7511d	Enforcement for Severe and Extreme ozone nonattainment areas for failure to attain	Sec. 185
7511e	Transitional areas	Sec. 185A
7511f	NO_x and VOC study	Sec. 185B
Subpart 3 -	Additional Provisions for Carbon Monoxide Nonattainment Areas	
7512	Classification and attainment dates	Sec. 186
7512a	Plan submissions and requirements	Sec. 187
Subpart 4 -	Additional Provisions for Particulate Matter Nonattainment Areas	
7513	Classifications and attainment dates	Sec. 188
7513a	Plan provisions and schedules for plan submissions	Sec. 189
7513b	Issuance of RACM and BACM guidance	Sec. 190
Subpart 5 -	Additional Provisions for Areas Designated Nonattainment for Sulfur Oxides, Nitrogen Dioxide, or Lead	
7514	Plan submission deadlines	Sec. 191
7514a	Attainment dates	Sec. 192
Subpart 6 -	Savings Provisions	
7515	General savings clause	Sec. 193
Subchapter II -	Emission Standards for Moving Sources	
Part A -	Motor Vehicle Emission and Fuel Standards	
7521	Emission standards for new motor vehicles or engines	Sec. 202
7522	Prohibited acts	Sec. 203
7523	Actions to restrain violations	Sec. 204
7524	Civil penalties	Sec. 205
7525	Motor vehicle and engines testing and certification	Sec. 206
7541	Compliance by vehicles and engines in actual use	Sec. 207
7542	Information collection	Sec. 208
7543	State standards	Sec. 209
7544	State grants	Sec. 210
7545	Regulation of fuels	Sec. 211

Table 3. (Continued)

42 U.S.C.	Section Title	Clean Air Act, as amended
7547	Nonroad engines and vehicles	Sec. 213
7548	Study of particulate emissions from motor vehicles	Sec. 214
7549	High altitude performance adjustments	Sec. 215
7550	Definitions	Sec. 216
7551	Study and report on fuel consumption of CAAA of 1977	Sec. 203
7552	Motor vehicle compliance program fees	Sec. 217
7553	Prohibition on production of engines requiring leaded gasoline	Sec. 218
7554	Urban bus standards	Sec. 219
Part B -	Aircraft Emissions Standards	
7571	Establishment of standards	Sec. 231
7572	Enforcement of standards	Sec. 232
7573	State standards and controls	Sec. 233
7574	Definitions	Sec. 234
Part C -	Clean Fuel Vehicles	
7581	Definitions	Sec. 241
7582	Requirements applicable to clean-fuel vehicles	Sec. 242
7583	Standards for light-duty clean-fuel vehicles	Sec. 243
7584	Administration and enforcement as per California standards	Sec. 244
7585	Standards for heavy-duty clean-fuel vehicles	Sec. 245
7586	Centrally fueled fleets	Sec. 246
7587	Vehicle conversions	Sec. 247
7588	Federal agency fleets	Sec. 248
7589	California pilot test program	Sec. 249
7590	General provisions	Sec. 250
Subchapter III -	General Provisions	
7601	Administration	Sec. 301
7602	Definitions	Sec. 302
7603	Emergency powers	Sec. 303
7604	Citizen suits	Sec. 304
7605	Representation in litigation	Sec. 305
7606	Federal procurement	Sec. 306
7607	Administrative proceedings and judicial review	Sec. 307
7608	Mandatory licensing	Sec. 308
7609	Policy review	Sec. 309

42 U.S.C.	Section Title	Clean Air Act, as amended
7610	Other authority	Sec. 310
7611	Records and audits	Sec. 311
7612	Economic impact analyses	Sec. 312
7614	Labor standards	Sec. 314
7615	Separability	Sec. 315
7616	Sewage treatment plants	Sec. 316
7617	Economic impact assessment	Sec. 317
7619	Air quality monitoring	Sec. 319
7620	Standardized air quality modeling	Sec. 320
7621	Employment effects	Sec. 321
7622	Employee protection	Sec. 322
7624	Cost of vapor recovery equipment	Sec. 323
7625	Vapor recovery for small business marketers of petroleum products	Sec. 324
7625-1	Exemptions for certain territories	Sec. 325
7625a	Statutory construction	Sec. 326
7626	Authorization of appropriations	Sec. 327
7627	Air pollution from Outer Continental Shelf activities	Sec. 328
Subchapter IV-A	Acid Deposition Control	
7651	Findings and purposes	Sec. 401
7651a	Definitions	Sec. 402
7651b	Sulfur dioxide allowance program for existing and new units	Sec. 403
7651c	Phase I sulfur dioxide requirements	Sec. 404
7651d	Phase II sulfur dioxide requirements	Sec. 405
7651e	Allowances for states with emissions rates at or below 0.80 lbs./mmBtu	Sec. 406
7651f	Nitrogen oxides emission reduction program	Sec. 407
7651g	Permits and compliance plans	Sec. 408
7651h	Repowered sources	Sec. 409
7651i	Election for additional sources	Sec. 410
7651j	Excess emissions penalty	Sec. 411
7651k	Monitoring, reporting, and recordkeeping requirements	Sec. 412
7651l	General compliance with other provisions	Sec. 413
7651m	Enforcement	Sec. 414
7651n	Clean coal technology regulatory incentives	Sec. 415
7651o	Contingency guarantee, auctions, reserve	Sec. 416
Subchapter V -	Permits	
7661	Definitions	Sec. 501

Table 3. (Continued)

42 U.S.C.	Section Title	Clean Air Act, as amended
7661a	Permit programs	Sec. 502
7661b	Permit applications	Sec. 503
7661c	Permit requirements and conditions	Sec. 504
7661d	Notification to administrator and contiguous states	Sec. 505
7661e	Other authorities	Sec. 506
7661f	Small business stationary source technical and environmental compliance assistance program	Sec. 507
Subchapter VI -	Stratospheric Ozone Protection	
7671	Definitions	Sec. 601
7671a	Listing of class I and class II substances	Sec. 602
7671b	Monitoring and reporting requirements	Sec. 603
7671c	Phase-out of production and consumption of class I substances	Sec. 604
7671d	Phase-out of production and consumption of class II substances	Sec. 605
7671e	Accelerated schedule	Sec. 606
7671f	Exchange authority	Sec. 607
7671g	National recycling and emission reduction program	Sec. 608
7671h	Servicing of motor vehicle air conditioners	Sec. 609
7671i	Nonessential products containing chlorofluorocarbons	Sec. 610
7671j	Labeling	Sec. 611
7671k	Safe alternatives policy	Sec. 612
7671l	Federal procurement	Sec. 613
7671m	Relationship to other laws	Sec. 614
7671n	Authority of Administrator	Sec. 615
7671o	Transfers among parties to Montreal Protocol	Sec. 616
7671p	International cooperation	Sec. 617
7671q	Miscellaneous provisions	Sec. 618
[29 U.S.C. 655]	Chemical Process Safety Management	Sec. 304 of CAA of 1990
[29 U.S.C. 1662e]	Clean Air Employment Transition Assistance	Sec.1101 of CAA of 1990

Note: This table shows only the major U.S. Code sections. For more detail and to determine when a section was added, the reader should consult the official printed version of the U.S. Code.

End Notes

[1] Unlike the other NAAQS pollutants, ozone is not directly emitted, but rather is formed in the atmosphere by the interaction of volatile organic compounds (VOCs) and nitrogen oxides (NOx) in the presence of sunlight. The control of ozone is, thus, based on regulating emissions of VOCs and NOx.

[2] EPA modified the ozone standard, specified in the statute as 0.12 parts per million (ppm) averaged over a 1-hour period, to 0.08 ppm averaged over an 8-hour period, through regulations promulgated in July 1997. In April 2004, the agency promulgated an implementation rule for the new 8-hour standard. Under this rule, the 1-hour standard was revoked as of June 15, 2005, and areas that had not yet attained it were converted to new classifications depending on their 8-hour concentration of ozone. As a result of court challenges, the ramifications of this conversion to the 8-hour standard are still unfolding, but in general the former 1-hour nonattainment areas remain subject to the controls specified for their 1-hour category. New nonattainment areas that did not exceed the 1-hour standard, but do violate the 8-hour standard, in general are subject to more flexible controls under Subpart 1 (Sections 171-179B) of the act. The standard was revised again in March 2008, to 0.075 ppm averaged over 8 hours, but EPA subsequently agreed to reconsider the 2008 standard, a task it expects to complete in 2011. Nonattainment areas for the new standard would be designated following its promulgation.

[3] This section of the report was written by Linda-Jo Schierow, Specialist in Environmental Policy.

[4] The 1990 amendments specified 189 pollutants, but P.L. 102-187, enacted on December 4, 1991, deleted hydrogen sulfide from the list of toxic pollutants, leaving only 188. On December 19, 2005, EPA removed methyl ethyl ketone (MEK) from the list of toxic air pollutants. The total number of listed air toxics is now 187.

[5] This section of the report was written by Larry B. Parker, Specialist in Energy and Environmental Policy.

[6] This section of the report was written by Larry B. Parker, Specialist in Energy and Environmental Policy and James E. McCarthy, Specialist in Environmental Policy.

[7] This section of the report was written by Larry B. Parker, Specialist in Energy and Environmental Policy.

[8] This section of the report was written by Claudia Copeland, Specialist in Environmental Policy.

[9] This section of the report was written by Larry B. Parker, Specialist in Energy and Environmental Policy.

INDEX

A

B

C

D

E

F

G

H

I

J

K

L

M

N

O

P

Q

R

S

V

W

Y